RATS, LICE AND HISTORY

BOOKS BY

HANS ZINSSER

RATS, LICE AND HISTORY
AS I REMEMBER HIM
The Biography of R.S.

RATS, LICE

AND

HISTORY

Being a Study in Biography, which, after Twelve
Preliminary Chapters Indispensable for the
Preparation of the Lay Reader, Deals
With the Life History of
TYPHUS FEVER

Also known, at various stages of its Adventurous Career, as *Morbus pulicaris*
(Cardanus, 1545); *Tabardiglio y puntos* (De Toro, 1574); *Pintas; Febris pur-
purea epidemica* (Coyttarus, 1578); *Febris quam lenticulas vel puncticulas vocant*
(Fracastorius, 1546); *Morbus hungaricus; La Pourpre; Pipercorn; Febris
petechialis vera; Febris maligna pestilens; Febris putrida et maligna; Typhus
carcerorum;* Jayl Fever; *Fièvre des hôpitaux; Pestis bellica; Morbus castrensis;*
Famine Fever; Irish Ague; *Typhus exanthematicus; Faulfieber; Hauptkrank-
heit; Pestartige Bräune; Exanthematisches Nervenfieber,* and so forth, and so forth.

By HANS ZINSSER

LITTLE, BROWN AND COMPANY
BOSTON   NEW YORK   TORONTO   LONDON

HC: 40 39 38 37 36 35
PB: 50 49 48 47 46 45 44 43 42 41

BP

PRINTED IN THE UNITED STATES OF AMERICA

This book is dedicated in affectionate friendship to Charles Nicolle, scientist, novelist, and philosopher

PREFACE

THESE chapters — we hesitate to call so rambling a performance a book — were written at odd moments as a relaxation from studies of typhus fever in the laboratory and in the field. In following infectious diseases about the world, one ends by regarding them as biological individuals which have lived through centuries, spanning many generations of men and having existences which, in their developments and wanderings, can be treated biographically. Typhus fever lends itself — more than most others — to such treatment because of its extraordinary parasitic cycles in the insect and animal worlds, the salient facts of which have all been elucidated within the last ten years. In no other infection does the bacteriologist find so favorable an opportunity for study of the evolution of a parasitism. Moreover, in its tragic relationship to mankind this disease is second to none — not even to plague or to cholera.

In the course of many years of preoccupation with infectious diseases, which has taken us alternately into the seats of biological warfare and into the laboratory, we have become increasingly impressed with the importance — almost entirely neglected by historians and sociologists — of the influence of these calamities upon the fate of nations indeed upon the rise and fall of civilizations.

The chapters which deal with this phase of our subject represent little more than preliminary notes. They may serve to stimulate future historians, who possess the learning which we lack, to give these factors the attention which they merit and to interpolate their effects into the interpretations of the past history of mankind.

In no sense can we claim to have made any original contributions to the history of medicine. We have taken information where we could find it, and have freely used the works of such profound scholars as Schnurrer, Hecker, Ozanam, Haeser, Hirsch, Murchison, and others. In consulting ancient and mediæval texts our meagre classical learning was reënforced by the charitable good nature of our colleagues Professors Gulick and Rand, of our friend Dr. Fred B. Lund, and by the enthusiastic interest of Mr. C. T. Murphy of the Harvard Classical Department. Conversation and correspondence with Professor Sigerist of Johns Hopkins, Professor Merriman of Harvard, Major Hume of the United States Army, and many others have brought us invaluable aid in critical places. We owe a particular debt of gratitude to our wise and kindly friend, Professor W. Morton Wheeler, who has been generous with advice and encouragement. Since this is, in no sense, a scientific treatise, we have left out references to recent work and, in order to neglect no one, have mentioned almost no names.

For our chapters and comments on matters of literary interest we make no apologies. Although we regard them as pertinent to the general scheme of our exposition, many will regard them as merely impertinent. But, in a way,

this book is a protest against the American attitude which
tends to insist that a specialist should have no interests
beyond his chosen field — unless it be golf, fishing, or
contract bridge. A specialist — in our national view —
should stick to his job like "a louse to a pig's back." We
risk — because of this performance — being thought less
of as a bacteriologist. It is worth the risk. But the day has
twenty-four hours; one can work but ten and sleep but
eight.

We hold that one type of intelligent occupation should,
in all but exceptional cases, increase the capacity for
comprehension in general; that it is an error to segre-
gate the minds of men into rigid guild classifications; and
that art and sciences have much in common and both may
profit by mutual appraisal. The Europeans have long appreciated this. That our book has contributed in this respect we have not the temerity to assert. At any rate, we
have written along as it has suited our fancy, and have
been amused and rested in so doing.

                                                    H. Z.

CONTENTS

RATS, LICE AND HISTORY

CHAPTER I

In the nature of an explanation and an apology

I

THIS book, if it is ever written, and — if written — it finds a publisher, and — if published — anyone reads it, will be recognized with some difficulty as a biography. We are living in an age of biography. We can no longer say with Carlyle that a well-written life is as rare as a well-spent one. Our bookstalls are filled with stories of the great and near-great of all ages, and each month's publishers' lists announce a new crop. The biographical form of writing has largely displaced the novel, it has poached upon the territory that was once spoken of as criticism, it has gone into successful competition with the detective story and the erotic memoir, and it has even entered the realm of the psychopathic clinic. One wonders what has released this deluge.

There are many possible answers. It is not unlikely that, together with other phases of modern life, literature has gone "scientific." As in science, a few men of originality work out the formulas for discovery in a chosen subject, and a mass of followers apply this formula to analogous problems and achieve profitable results. In an age of meagre literary originality, it is a natural impulse for workers to endeavor to explain the genius of great masters. And for every novelist, poet, or inventor of any kind, we have a dozen interpreters, commentators, and critics.

Once biography was a serious business and the task of the scholar. When Plutarch wrote his *Parallel Lives,* his mind — as Mr. Clough rightly remarks — was running on the Aristotelian ethics and the Platonic theories which formed the religion of the educated men of his time. He dealt less with action, more with motives and the reaction of ability and character upon the circumstances of the great civilizations of Greece and of Rome. Scholarly biographies of later ages followed similar methods, even in such intensely personal records as Boswell's *Johnson,* or the *Conversations* by which so dull an ass as Eckermann managed to write himself into permanent fame. The minor details of intimate life were, in the past, regarded as having consequence only as they had bearing on the states of mind that led to high achievement. It was recognized that *"les petitesses de la vie privée peuvent s'allier avec l'héroïsme de la vie publique."* But they were utilized only when they were significant or amusing. But all this has changed. The new school sees the key to personality in the *petitesses.* Biography has become neurosis-conscious. Freud is a great man. But it is dangerous when a great man is too easily half-understood. The Freudian high explosives have been worked into firecrackers for the simple to burn their fingers. It has become easy to make a noise and a bad smell with materials compounded by the great discoverer for the blasting of tunnels. Biography is obviously the best playground for the dilettante of psychoanalysis. The older biographers lacked this knot-hole into the subconscious. They judged

their heroes only by the conscious. The subconscious de-
thrones the conscious. Great men are being reappraised
by their endocrine balances rather than by their perform-
ances. Poor Shelley! Poor Byron! Poor Wagner! Poor
Chopin! Poor Heine! Poor Mark Twain! Poor Henry
James! Poor Melville! Poor Dostoevski! Poor Tolstoy!
And even poor Jesus! There are still a lot left — the
surface is hardly scratched. But even before the great ones
give out, the "damaged" ones make good reading: P. T.
Barnum, Brigham Young — even unto Al Capone and
Pancho Villa.

In the present biography, we are forced by the nature of
our subject to revert to the older methods. We will profit
by no assistance from psychoanalysis. There will be no
prenatal influences; no Œdipus or mother complexes; no
early love affairs or later infidelities; no perversions,
urges, or maladjustments; no inhibitions by respectability,
and no frustration by suppressed desires. We shall have
no gossip to help us; no personal letters which there was
no time to burn. We cannot count upon the *réclame* of a
libel suit barely averted, or of scandals deftly hinted at.
We have not even the comfort of preceding biographers
and essayists whom we can copy, paraphrase, or refute.
Indeed, we are quite stripped of the sauces, spices, and
dressings by which biographers can usually make poets
and scientists into quite ordinary and often objectionable
people; by which they can divert attention from the work
of a man to his petty or perhaps vicious habits; by which
they can create a hero out of a successful commercial
highbinder; by which they can smother public guilt by

domestic virtue, or direct interest from the best and lasting accomplishments of their subject to the utterly unimportant private matters of which he was ashamed.

The habitué of biographies will ask himself how, without these indispensable accessories of the biographical tradesman, we can dare to enter this field. The answer is a simple one: the subject of our biography is a disease.

We shall try to write it in as untechnical a manner as is consistent with accuracy. It will of necessity be incomplete, for the life of our subject has been a long and turbulent one from which we can select only the high spots. Much of its daily domestic history has been as commonplace and repetitive as that of any human being, warrior, poet, or shopkeeper. Above all, our narrative is not "popular science." If our story is, in places, dramatic, it will be the fault of the story — not our own. Nobody will be educated by it. We have chosen to write the biography of our disease because we love it platonically — as Amy Lowell loved Keats — and have sought its acquaintance wherever we could find it. And in this growing intimacy we have become increasingly impressed with the influence that this and other infectious diseases, which span — in their protoplasmic continuities — the entire history of mankind, have had upon the fates of men.

In approaching our subject, however, we permit ourselves a number of digressions into which our undertaking inevitably forces us.

2

Infectious disease is one of the great tragedies of living things — the struggle for existence between different forms of life. Man sees it from his own prejudiced point of view; but clams, oysters, insects, fish, flowers, tobacco, potatoes, tomatoes, fruit, shrubs, trees, have their own varieties of smallpox, measles, cancer, or tuberculosis. Incessantly, the pitiless war goes on, without quarter or armistice — a nationalism of species against species. Usually, however, among the so-called "lower" forms of life, there is a solidarity of class relationship which prevents them from preying upon their own kind by that excess of ferocity which appears to prevail only among human beings, rats, and some of the more savage varieties of fish. There are, it must be admitted, isolated instances in the animal kingdom of a degree of ferocity within the same species not yet attained by man. Husband eating is an accepted custom with the spiders, and among the Alacran or Scorpions, it is quite *de rigueur* for the mother to devour the father and then, in her turn, to be eaten by her "kiddies." When male members of the larger cat families — that is, mountain lions — waylay and eat their own children, this is not truly an evidence of ferocity. It is an indirect *crime passionnel;* the result of an impatient tenderness for the lioness who has become too exclusively the mother. The motive is love, and, as La Rochefoucauld has said, "*Si on juge l'amour par la plupart de ses effets, il ressemble plus à la haine qu'à l'amitié.*"

Nature seems to have intended that her creatures feed upon one another. At any rate, she has so designed her cycles that the only forms of life that are parasitic directly upon Mother Earth herself are a proportion of the vegetable kingdom that dig their roots into the sod for its nitrogenous juices and spread their broad chlorophyllic leaves to the sun and air. But these — unless too unpalatable or poisonous — are devoured by the beasts and by man; and the latter, in their turn, by other beasts and by bacteria. In the Garden of Eden perhaps things may have been so ordered that this mutual devouring was postponed until death, by the natural course of old age, had returned each creature's store of nutriment to the general stock. Chemically, this might have been possible, and life maintained. But in the imperfect development of cohabitation on a crowded planet, the habit of eating one another — dead and alive — has become a general custom, instinctively and dispassionately indulged in. There is probably as little conscious cruelty in the lion that devours a missionary as there is in the kind-hearted old gentleman who dines upon a chicken pie, or in the staphylococcus that is raising a boil on the old gentleman's neck. Broadly speaking, the lion is parasitic on the missionary, as the old gentleman is on the chicken pie, and the staphylococcus on the old gentleman. We shall not enlarge upon this, because it would lead us into that excess of technicality which we wish to avoid.

The important point is that infectious disease is merely a disagreeable instance of a widely prevalent tendency of all living creatures to save themselves the bother of building, by their own efforts, the things they require. Whenever they find it possible to take advantage of the constructive labors of others, this is the direction of the least resistance. The plant does the work with its roots and its green leaves. The cow eats the plant. Man eats both of them; and bacteria (or investment bankers) eat the man. Complete elucidation would require elaborate technical discussions, but the principle is clear. Life on earth is an endless chain of parasitism which would soon lead to the complete annihilation of all living beings unless the incorruptible workers of the vegetable kingdom constantly renewed the supply of suitable nitrogen and carbon compounds which other living things can filch. It is a topic that might lend itself to endless trite moralizing. In the last analysis, man may be defined as a parasite on a vegetable.

That form of parasitism which we call infection is as old as animal and vegetable life. In a later chapter we may have occasion to consider its origin; to this we have some clue from the new diseases which appear to be constantly developing as we begin to conquer the old ones. But our chief purpose in writing the biography of one of these diseases is to impress the fact that we are dealing with a phase of man's history on earth which has received too little attention from poets, artists, and historians. Swords and lances, arrows, machine guns, and even high explosives have had far less power over the fates of the nations than the typhus louse, the plague flea, and the yellow-fever mosquito. Civilizations have retreated from the plasmodium of malaria, and armies have crumbled into rabbles under the onslaught of cholera spirilla, or of dysentery and typhoid bacilli. Huge areas have been devastated by the trypanosome that travels on the wings of the tsetse fly, and generations have been harassed by the syphilis of a courtier. War and conquest and that herd existence which is an accompaniment of what we call civilization have merely set the stage for these more powerful agents of human tragedy.

3

Having written the preceding paragraphs, we read them over and came to the conclusion that there was little in them that mattered very much. We were, perhaps, a little severe in discussing modern biographers. One is lured into discussions of this kind by one's irritations. One can disagree with many of the opinions expressed by Goethe in Eckermann, or by Renan, or Sainte-Beuve, or by Babbitt, or by Whitehead, — when one understands what he is talking about, — and come away with the satisfaction of having been stimulated to oppose views by the importance of those one disagreed with. But one is merely irritated by the complacency with which the sciences and the arts are dealt with *e superiore loco* by the younger school of American biographical critics, who sit between intelligence and beauty, — like Voltaire between Madame de Staël and a flirtatious Marquise, — "without possessing either." One wishes to exclaim, with a similarly irritated Frenchman: "Save us, dear Lord, from the *connaisseurs qui n'ont pas de connaissance* and from the *amateurs qui n'ont pas d'amour!*" A part of our first chapter, therefore, is nothing more than a growl. Yet it still serves to introduce our subject; and we are further inclined to retain it for the following reasons. We are engaged in an occupation which philosophers, mathematicians, physicists, physical chemists, biochemists, and even physiologists (who may in many cases have been less valuable to science than one of Pawlow's dogs) do not acknowledge as a science; and which poets, novelists, critics, biographers, dramatists, painters, sculptors, and even journalists categorically exclude from the arts. We are in a position, therefore, to look both ways with the clarity begotten of humility. But, in discussing our ideas with representatives of the various callings named above, we encountered a common misconception — perhaps the only opinion on which there was agreement — to the effect that men were impelled to enter the career of investigating infectious diseases from a noble desire to serve mankind, to save life, and to relieve suffering.

A friend of ours is a professional writer. By this, we mean a person who makes his living by writing in the same way that a bricklayer makes his by laying bricks, or a plumber supports himself by sweating joints. Writing, of course, like speech, is a method of expressing ideas or telling tales. It is also a means of conveying to others emotions, conceptions, or original comprehensions which might instruct, amuse, delight, or elevate. This kind of writing used to be called art. And once — when only the intelligent could read — writing also needed to be intelligent and artistic.

In our day, however, all kinds of people can read:

college professors and scrubwomen, doctors and lawyers, bartenders, ministers of the gospel and trained nurses. They all have the same ideal of the happy ending of a dull day — a comfortable couch, a bed lamp, and something to read. And there must, in consequence, be writers to supply this need — literature for the intelligent as for the moron — a book for every brain, like a motor car for every purse.

The particular writer of whom we speak has been unusually successful in alternately supplying both markets — now satisfying the reasonably intelligent, and again luring a fat check with stories about the poor boy and the boss's daughter. In the latter mood, he has scented the rich possibilities of exploiting the sensationalisms of science — a source of revenue so successfully tapped by a number of his literary contemporaries. But never having had any close association with workers in the field of infectious diseases, he shared this misconception of the noble motives which impelled these queer people. And not quite understanding how anyone could be impelled by noble motives, he asked us: "How do bacteriologists get that way?" We answered his question more or less in the following manner.

A great deal of sentimental bosh has been written about this totally erroneous assumption. When a bacteriologist dies — as other people do — of incidental dissipation, accident, or old age, devotion and self-sacrifice are the themes of the minister's eulogy. Let him succumb in the course of his work, — as an engineer falls down a hole, or a lawyer gets shot by a client, — he is consecrated as a martyr. Novelists use him as they formerly did cavalry officers, Polish patriots, or aviators. If an epidemiologist on a plague study talked and behaved in the manner of the hero of *Arrowsmith*, he would not only be useless, but he would be regarded as something of a yellow ass and a nuisance by his associates. And de Kruif is far too intelligent a man not to have known, when he wrote his thriller on *Men against Death*, that raucous laughter would be its reception in the laboratories and in the field where the work he describes is being done.

As a matter of fact, men go into this branch of work from a number of motives, the last of which is a self-conscious desire to do good. The point is that it remains one of the few sporting propositions left for individuals who feel the need of a certain amount of excitement. Infectious disease is one of the few genuine adventures left in the world. The dragons are all dead, and the lance grows rusty in the chimney corner. Wars are exercises in ballistics, chemical ingenuity, administration, hard physical labor, and long-distance mass murder. Ships have wireless equipment. Our own continent is a stage route of gas stations, and the Indians own oil wells. Africa is a playground for animal photographers or museum administrators and their wives, who go there partly to have their pictures taken with one foot on a dead lion or elephant and disgusted-looking black boys carrying boxes of champagne and biscuits on their patient heads. Flying is adventurous enough, but little more than a kind of acrobatics for garage mechanics, like automobile racing. But however secure and well-regulated civilized life may become, bacteria, Protozoa, viruses, infected fleas, lice, ticks, mosquitoes, and bedbugs will always lurk in the shadows ready to pounce when neglect, poverty, famine, or war lets down the defenses. And even in normal times they prey on the weak, the very young and the very old, living along with us, in mysterious obscurity waiting their opportunities. About the only genuine sporting proposition that remains unimpaired by the relentless domestication of a once free-living human species is the war against these ferocious little fellow creatures, which lurk in the dark corners and stalk us in the bodies of rats, mice, and all kinds of domestic animals; which fly and crawl with the insects, and waylay us in our food and drink and even in our love.

CHAPTER II

Being a discussion of the relationship between science and art — a subject that has nothing to do with typhus fever, but was forced upon us by the literary gentleman spoken of in the last chapter

I

This chapter will be received with contemptuous shrugs by the professionally literary. There is a prejudice in America that specialists should not trespass beyond their own paddocks, however interestedly they may look over the rails. But literary critics are constantly telling us that science is this or that — "science should not be exalted out of its place," and so on; and since we cannot possibly know less about literature than most of these gentlemen know about science, we venture to proceed, hoping that Messrs. Edmund Wilson, Van Wyck Brooks, Mumford, Max Eastman, and others who were the "Younger School," until they grew middle-aged, will skip this part of our book.

The biologist is in a peculiarly difficult position. He cannot isolate individual reactions and study them one by one, as the chemist often can. He is deprived of the mathematical forecasts by which the physicist can so frequently guide his experimental efforts. Nature sets the conditions under which the biologist works, and he must accept her terms or give up the task altogether.

He knows that physicochemical analysis will never give the final clue to life processes; yet he recognizes that "vitalism" and "neovitalism" are little more than a sort of amorphous theology born of a sense of the helplessness of mere "mechanism."[1] So the patient biologist plods along, piling up his empirical observations as honestly as he can — getting what satisfaction he may from the fact that he is helping, by infinite increments, to reduce the scope of vitalistic vagueness to narrower and narrower limits. As Bergson puts it: "A very small element of a curve is near being a straight line; and the smaller it is, the nearer. . . . So, likewise, 'vitality' is tangent, at any and every point, to physical and chemical forces. . . . In reality [however], life is no more made up of physico-chemical elements than a curve is composed of straight lines." The biologist is constantly differentiating the curve of vitality, quite aware that mankind can approach, but never reach, the "limiting value" of complete comprehension. Moreover, he knows — whenever he attacks a problem — that before he can advance toward his objective, he must first recede into analysis of the individual elements that compose the complex systems with which he is occupied.

Such difficulties engender a habit of mind that has hampered us in the present undertaking. We approached the writing of the biography of typhus fever with the care-

---

[1] And, indeed, ultimately they both encounter the same inevitable perplexity, since, as Paley rightly asserts, mechanism presupposes God as the mechanician. This is the difficulty faced by all the recent astronomical and physicist school of ponderers.

less confidence which always accompanies the first conception of an experimental objective. We were first deflected into contemplation of the general methods of biographical writing; then arose the question why men occupied themselves with the study of disease. We thought we were through with preliminaries, when our literary friend dropped in again, and proceeded to scatter salt upon our enthusiasm.

"How," he said, "can a person who has spent his life cultivating bacteria; inoculating guinea pigs, rabbits, mice, horses, and monkeys; posting about the dirty corners of the world in the study of epidemics; catching rats in foreign cellars; disinfecting, delousing, fumigating; looking at rashes, down throats and into other apertures of man and animals; breeding lice, bedbugs, fleas, and ticks; examining sputum, blood, urine, stools, milk, water, and sewage — how," he repeated, "can such a person, who is not quite a scientist and nothing of an artist, presume to undertake a task which no one not an artist could successfully accomplish? You might be right about the keyhole biographers and the pasteurized Rabelaisian school of Freudian critics, but is that any worse than the literary-scientific spinster movement? Do you want to be like Dr. Collins of New York, 'the-Doctor-looks-at-this, the-Doctor-looks-at-that' business?"

"But!" we replied —

"Look at all the rest of the middle-aged scientists who have made fools of themselves dabbling with art. Read the *Atlantic Monthly*."

"Good Lord," I said, "one need n't stop being a bacteriologist just because one takes an intelligent interest in other things. Here in America we seem to expect a specialist to become a sort of Taylorized factory worker. Why should a man look at the world through only one knot-hole?"

"Oh, look through a dozen or climb up and look over the fence if you like. But keep still about things you 're not trained to handle. Biography is a job for an artist. Stick your head out of your laboratory window and watch the world go by. But if you want to write, pull it in again and write for the *Journal of Experimental Medicine.* You 'll only end, if you keep this up, by losing what little reputation you 've got."

"But," we demurred, "is a man to be denied an intelligent appreciation of art just because he knows something about a science? Is literature to be appraised only by those who have time to read after breakfast? What 's the essential difference between art and science anyway?"

"That 's a difficult question," he said. "Goethe might have answered it, but he did n't think it was worth while. The late war between humanists and antihumanists might have brought an answer — only both sides were so angry at each other and so ignorant of science that they neglected the main issue. Babbitt, with his vast erudition, might have found a reply if he had lived. Toward the end, the small fry were keeping him too busy with his heels. Anyway, neither you nor I know enough to deal with it."

Our friend's opinions on matters of this sort have always carried much weight with us, and, in this case, they impelled us to delay embarking upon our project — which, as he said, transcended our scientific t..ining — until we had given thought to the essential differences, if there were any, between science and art.

We approached the problem modestly by examining the opinions of others, and found that men far wiser than ourselves had failed to agree. Eddington and Jeans incline to limit science to the "metrical or mathematical descriptions of phenomena," a conception which would exclude even the biological branches of learning. But having ascended to these cold heights by laborious upward paths of reason, they sit down in their metaphysical toboggans and swish back into the warm and comfortable vales of theology. Dingle attempts a more liberal view, defining science as a method of "dealing rationally with experiences which have a certain quality; namely, that they are common to all normal people." This is dreadful English, but — once parsed — it means, conversely, that the territory of art is that of experiences which are "peculiar to the individual, or perhaps shared by a limited number of others." This opinion is much like the pre-Darwinian method of classifying animals by their superficial similarities, which made the whale a fish and the bat a bird. Whitehead penetrates more deeply beneath the mere morphology of the problem into its comparative anatomy and physiology. He includes, in the category of science, the biological branches and geology, and, more than that, he regards naturalistic art (Leonardo) as closely akin to science. Indeed, he finds in great literatures — for instance, in the "scientific imagination" of Æschylus,

Sophocles, and Euripides, in their visions of Fate "urging a tragic incident to its inevitable issue" — the same principle of "Order" which is the "vision possessed by science." If Aristotle could return to us long enough to familiarize himself with modern scientific thought, we venture to say he would come pretty close to agreeing with Whitehead. Incidentally, what a kick Aristotle would get out of Harvard!

That any sharp separation between science and art is impossible was also in the mind of Havelock Ellis, when he wrote the following passage: "To press through, to reveal, to possess, to direct and to ennoble, that is the task and the longing alike of the lover and the natural discoverer; so that every Ross or Franklin is a Werther of the Pole, and whoever is in love is a Mungo Park of the spirit." We should have taken more pleasure in this quotation had Mr. Park's Christian name been other than "Mungo." But, as it stands, it expresses the burden of the thought that was developing in our mind.

2

To most of the modern literary critics — probably because of their almost incredible ignorance of scientific thought — the so-called scientist is a "mere rationalist," and science is held, in respect to art, as photography is to painting. This separation on the basis of precision is utterly untenable. Science is not a whit more photographic than is art. Measurements and formulations are, even in the so-called exact — the physical — sciences, not much more than reasonably accurate approximations. Scientific method is again and again forced to employ abstract conceptions, irrational numbers like $\sqrt{2}$ and $\sqrt{3}$, the line without breadth, the point without volume, zero, the negative quantity, or the idea of infinity. And scientific thought continually sets sail from ports of hypothesis and fiction,[2] advance bases of the exploring intellect. Matter becomes molecules, molecules become atoms; atoms, ions; ions, electrons; and these, in turn, become uncomprehended sources of energy — not more clear as seizable reality than the poet's conception of the "soul," which he knows only from its "energy" — the yearnings, delights, and sorrows which he feels. The history of science is full of examples of what, in art, would be spoken of as inspiration, but for which Whitehead's definition, "speculative reason," seems much more appropriate.

It is only too painfully obvious, moreover, that neither the scientist nor the artist is ever a "creator." The word "creative," so incessantly misused by our younger critical schools, is a fiction of that optimism about human achievement which — it has been said — thrives most vigorously in lunatic asylums. Nature, as Goethe puts it, runs its course by such eternal and necessary principles that even the gods themselves cannot alter them. The most that the scientist and the artist accomplish is new understanding of things that have always been. They "create" a clearer perception. They are both, in this sense, observers, the obvious difference being that the scientist impersonally describes the external world, whereas the artist expresses

[2] This has been clearly set forth in Hans Vaihinger's *Die Philosophie des Als Ob.*

the effects which external things exert upon his own mind and heart. In both cases, the more generally applicable the observations, the greater is the science or art.[3]

Would it not be fair to say that an achievement of observation becomes science or art according to the degree to which its comprehension calls upon perception by the reason or by the emotions, respectively? The capacities of intelligence form a sort of spectrum which extends from what we may call an infra-emotional to an ultra-reason range. At the infra-emotional extreme lie the perceptions set in motion by music and by lyrical poetry. At the opposite end — that of pure reason — is placed the perceptional capacity for mathematics. Between the two there is a wide range of overlapping where art is scientific and science artistic. Literature in the sense of prose may be taken to hold a middle ground, shading on the left into epic and narrative poetry, and on the right through psychology, biology, and so forth, toward mathematics.

"What happens when you go off the deep end of either side?" asked my friend.

[3] I. A. Richards expresses this function of the artist as an observer of the "facts" of human emotions in a precise manner when he says, "In the arts we find the record, in the only form in which these things can be recorded, of the experiences which have seemed worth having to the most sensitive and discriminating persons." In this sense Leonardo, Shakespeare, Cervantes, Goethe, Dostoevski, and countless other artists were as truly accurate observers in the field of human experiences as were Newton and Pascal in the field of the external world.

André Gide means the same thing when he says, "Everything has always existed in man . . . and what new times uncover in him has always slumbered there. . . . How many hidden heroes await only the example of a hero in a book, only a spark of life given off by his life in order to love, only a word from him in order to speak."

"Well, beyond the 10^{-10} range experience seems to show that the end organs give out and the physicist joins the church; whereas on the other side, as I should judge from Joyce, Gertrude Stein, and their imitators, the spinal cord begins to horn in on the brain. In either case it ceases to be science or art."

3

I continued the discussion with my friend at our next meeting.

"On that basis," he said, "it should be easy to classify any performance by a sort of intellectual spectroscopic analysis."

"With the older forms it was usually easy to fit them into their proper places in the spectrum. Critics like Coleridge or Sainte-Beuve needed to concern themselves only with style, beauty of diction, clarity of thought, intensity, sincerity, depth, and the qualities of taste and sensitiveness which, while vague and subtle, were still within the scope of the underanged mind. Art could be judged by any informed and intelligent critic without recourse to borderline psychiatry. The corner was turned by the French symbolists — who followed Baudelaire, Rimbaud, Verlaine, Mallarmé, Laforgue. On occasion these great men came close to the jumping-off place of uncomprehensibility. But in the main they achieved a great beauty by the very dusk and mist through which their thoughts, sufferings, and joys were mysteriously, grotesquely, vaguely, but still effectively perceived. One cannot, with Lasserre, deny them their just places merely because they applied their superb gifts to *tristesse* and *laideur*. We make no plea for a return to Tennysonism or the Longfellow era, but had Sainte-Beuve been required to pass judgment on certain passages of T. S. Eliot, the later Joyce, or Gertrude Stein, he would surely have gone into consultation with Charcot or Bernheim, a dilemma which our modern critics seem to admit — in their judgments of modern work — by their habitual appeal to Sigmund Freud. It is, of course, difficult, even in medical practice, to survey sharply the line between sanity and borderline derangements. But when the critic of a work of art needs psychiatric training, this fact alone would serve to throw suspicion on the artistic value of his subject. The real difficulty of applying our kind of spectroscopic analysis to much of the modern stuff lies in the fact that a good deal of it lacks the rationality of science without possessing the emotional appeal of art.

"Let us examine some of it. Take T. S. Eliot — who, in his prose, shows great clarity of thought and to whom no one will deny talent, originality, and, on occasion, great beauty. But in much of his poetry he plays, as has been aptly remarked, a guessing game with readers, whom he seems to appraise, apparently with some reason, as imbeciles. 'Guess which memory picture of my obviously one-sided erudition I am alluding to? See note 6a.' Then he drops suddenly, after a few lines of majestic verse, into completely irrelevant babble.

"In the room the women come and go
Talking of Michelangelo."

One is tempted to add, 'Eenie, meenie, minie, mo.' Or
this: —

> "Madame Sosostris, famous clairvoyante,
> Had a bad cold, nevertheless
> Is known to be the wisest woman in Europe
> With a wicked pack of cards.

"Why 'nevertheless'? Was she wise *because* she had a
bad cold? Or this (one has the choice of innumerable passages): —

> "Now Albert's coming back, make yourself a bit smart.
> He 'll want to know what you done with that money he gave you
> To get yourself some teeth.

"Is that poetry? It sounds like trivial prose. It certainly
is n't science."

"Of course it 's not fair to take things out of their con-
texts like that. The thing as a whole symbolizes the Waste
Land of modern disillusionment. Of course it 's hard for
a scientist to understand."

"It 's not whether a thing is hard to understand. It 's
whether, once understood, it makes any sense. Every now
and then my monkeys get loose in the laboratory and
achieve brilliant and bizarre effects by smashing bottles
of colored liquids against microscopes and Bunsen burners.
The result is a stimulating chaos of lights, sounds, and
excitements. But when they get through there 's nothing
left but disorder and litter that has to be swept up before
orderly scientific work can be resumed. You can do the
same thing with the workshops of art. What I don't under-

stand is why a man of such obvious power will do that
sort of thing."

"I suppose you will say the same thing about Baudelaire?" he said.

"Oh, dear, it 's the old stuff that these people derive
themselves from Baudelaire and Rimbaud and Laforgue.
But those men were making discoveries. Baudelaire was
an organic chemist. He synthetized extraordinarily re-
pulsive but new compounds. But incoherence and a bad
smell don't make a Baudelaire."

"Well, let's try another; perhaps you recognize this
one?

"Nearly all of it to be as a wife has a cow.   All of it to be as a
wife has a cow, all of it to be as a wife has a cow, a love story.
As to be all of it as to be a wife as a wife has a cow, a love story,
all of it as to be all of it as a wife all of it as to be as a wife has a
cow a love story. . . .

or

"A meal is mutton mutton why is lamb cheaper, it is cheaper
because so little is more."

"That 's Gertrude Stein," I said, "but listen to this
one: —

"Balloons — colored balloons — my colored balloons — Who
busted my balloons? Bolony balloons; they have punctured my
categorical imperative."

"I don't seem to remember that in her writings," he
replied.

"No, that is n't Gertrude Stein. That 's Alice Gray,
whom I knew in the McLean Hospital. She was fifty,
but she imagined she was a baby. Listen to another: —

"Pease porridge hot, pease porridge cold,
  Pease porridge in the pot . . ."

"You 're only trying to be funny," he interrupted me. "As a matter of fact, Gertrude Stein can write quite sensibly when she wants to."

"Why does n't she?" I asked.

"She 's practising automatic writing." [4]

"Then it 's science."

"Oh, no — she is creating an impression by an alternation of conscious and subconscious explosions."

"Then it 's art — in the sense of fireworks."

"But she 's had an immense influence on younger writers," he said.

"So have Mrs. Eddy and P. T. Barnum," I replied. "Without Baudelaire there might not have been a Rimbaud or a Verlaine. Without Buffalo Bill, P. T. Barnum, or Mrs. Eddy, there might have been no Gertrude Stein, and Joyce might have continued to write distinguished prose."

"Speaking of Joyce," he said, "have you tried 'Tam and Shem' or whatever their names are? Listen!

"Eins within a space and a weary wide space it wast, are wohned a Mookse. The onesomeness wast all to lonely, archunsitlike, broady oval and a Mookse he would a walking go (My hood! cries Antony Romeo). So one grand summer evening after a great morning and his good supper of gammon and spittish, having flabelled his eyes, pilleoled his nostrils, vacticanated his ears . . ."

"Stop!" I cried. "I got a licking for that sort of thing when I was a little boy."

[4] See B. F. Skinner in the *Atlantic Monthly* for January 1934.

"Is it science or art?" he asked.

"Neither, of course," I said. "But what puzzles me still is why they do it. It would be too easy to dismiss the matter by assuming that they were mildly crazy. Moreover, the ability of the ones we have mentioned to return, at will, to the rational state excludes this."

"You forget," he said, "the idea of Poesie Pure — the less it means, the better; the approximation of poetry to music of Walter Pater and of Moore."

"The relationship of poetry to music has also come in for a great deal of learned twaddle. Valéry says the poet is merely a sort of musician. Wyndham Lewis calls it 'critical mysticism.' They speak a lot (Brémond) about the 'summons from within,' the 'weight of immortality upon the heart,' poetry which 'goes further than the word which expresses it,' and so forth. Sometimes the critic goes much farther in his mysticism than the poets he writes about."

Incidentally it is a curious phenomenon that some of the great scientists when they become critics, and are caught in efforts to explain their own æsthetic reactions to poetry, become almost as mystical as the literary analysts. Occasionally a man's authority is so great — in most particulars rightly so — that to criticize him is, in the eyes of the learned world, like spelling God with a small *g*. I refer to Whitehead, and in disagreeing with him I feel much like a Neanderthal man attacking a mastodon with a bean-shooter. When he discusses the application of Clerk Maxwell's equation to the interior of the atom, he has me on my back. But when he begins to attribute reference to some form of Kantian, Berkeleyan, or Platonic idealism to Shelley in his poem on Mont Blanc, or derives Wordsworth's nature worship from a "criticism of science," he merely reveals his own inability to take his foot off the brake of reason and coast freely with the emotions.

Now, when Shelley writes about the cloud or about Mont Blanc, he is not thinking of the "elusive endless change of things," nor is he consciously refusing "to accept the abstract materialism of science." He is expressing in magnificent images the thoughts and emotions that are aroused in him by the nature he views; and no amount of philosophical analysis can give the reader Shelley's full effect. The sheer beauty of the shifting thoughts and feelings, and the musical beauty, — not only musical in sound, but in the harmony of images as well, — must arouse in the reader the same reaction, transmitted from the poet, which nature aroused in the poet himself. It is the old question that Shelley himself answered by saying: "To analyze a work of art into its elements is as useless as throwing a violet into a crucible." Of course, poetry approaches music, but unlike music it has the power of concreteness in thought and imagery. The greatest poetry is communication and is clear. It may, through pure lyricism, progress sanely to the symbolism of Mallarmé and his contemporaries, growing less and less intellectually clear — more and more dependent upon imagery and suggestion. When it goes beyond that, it may come to the deep end where it tries to be purely saxophonic, as in the "jug, jug, jug" or the "bam boo bim bam tree" gibberish in certain passages of Mr. Eliot. Baudelaire had this in mind when in *L'Art Romantique* he said that "there are subjects which belong to painting, others to music, others to literature," and *"Est-ce par une fatalité des décadences qu'aujourd'hui chaque art manifeste l'envie d'empiéter sur l'art voisin?"* [5] When a work of literature, even if it is written in short, capitalized lines, becomes utterly incomprehensible to the sane and sensitive, it has gone off the deep end.

Why, we must ask ourselves, have individuals of unquestionably great powers chosen to play with their minds like captive monkeys with their genitalia? It would be merely tragic had they not created a sort of "holy-roller" school of followers among the permanent intellectual undergraduates. Wyndham Lewis comes close to a definition when he calls it the "idiot child" cult — the child overshadowed by the imbecile. As we have said, Skinner thinks, in the Stein case, it is conscious experimentation with "automatic writing."

One could also postulate: —

(1) That they are consciously pulling the legs of the

---

[5] It is pertinent, in this connection, to ask oneself what would have been the result if D. H. Lawrence had been a professional instead of an occasional painter. A painted Lady Chatterley — the most exquisite technique notwithstanding — would surely have been so completely out of drawing, with the lower parts so much larger than the upper, as to have been hardly recognizable as a human figure. The picture could not have been hung, even in a speak-easy.

In this matter of disproportionate emphasis on those phases of a subject which correspond to the writer's own neuroses, literature can "get away" with a great deal that would be impossible in architecture, sculpture, painting, or even music.

large neo-intellectual public either for fun or for profit.

(2) That they are suffering from a well-recognized form of exhibitionism — the craving for sensational notice, whether approval or attack. This is the mild derangement that probably explains mediums. It is the impulse that, in a less pronounced form, leads people to write to the newspaper, to lend their names to cigarette advertisements, or to say in print that they "suffered from fits" until they had taken one bottle of Neuropop.

(3) That they are seriously carrying on psychological experiments with themselves — in which case, they ought to do it in decent privacy, as though they were taking drugs.

Or (4) that it is barely possible they are yielding to the uncontrollable impulse to expose their own diseases, just as the physically sick like to tell about their operations or their chronic colitis.

If they were commonplace people this exercise would attract only sympathetic attention. These are formidable machines and one wishes the insulation had not burnt off the power lines.[6]

However one looks at it, it appears to the medically informed that these people are substituting the spinal cord for the brain, or at any rate are moving down from the frontal lobes towards the basal ganglia.

[6] One could of course multiply examples with "cummins," Ezra Pound, and so forth. We distinctly exclude Hart Crane, whom we had occasion to know when we were working on typhus in Mexico. He was a man of great talent, appealing and tragic, for he was very sick in spirit.

"You 've talked a great deal," said my friend, "but in the end it comes down to a definition of beauty — does n't it?"

"Well, give me one," I replied.

"Here 's the latest one," he said. "Beauty is the mutual adaptation of the several factors in an occasion of experience. Thus in its primary sense, beauty is a quality which finds its exemplification in actual occasions. Or, put it conversely, it is a quality in which such occasions can severally participate."

"Hail to thee, blithe spirit," I replied. "Bird thou never wert."

"Well, let 's go on," he replied. "In order to understand this definition of beauty, it is necessary to keep in mind three doctrines which belong to the metaphysical system in terms of which the world is being interpreted in these chapters. These three doctrines, respectively, have regard to mutual relations (*a*) between the objective content of a prehension and the subjective form of that prehension, and (*b*) between the subjective form of various prehensions in the same occasion, and (*c*) between the subjective form of a prehension and the spontaneity involved in the subjective aim of the prehending occasion."

"Stop," I said. "Is that by Gertrude Stein?"

"No," he replied, "it 's by Whitehead."

"Well, I 'll be damned," I said. "I think I 've decided that it 's perfectly safe for me to go ahead with my biography of typhus."

Indeed, I reflected when my friend had departed, whenever I think about these things for any length of time I feel grateful for good honest diseases like typhus, syphilis, and a few others. You always know where you have them. And if you begin indulging in "whimso-whamso" while you are engaged with them they are sure to make a fool of you by putting you on your back. You either leave them alone or approach them with cautious competence. Think what might happen to our modern critics if the great dead whom they inexpertly dissect could infect them with psychic boils and carbuncles; or if Mr. Joyce's preoccupation with the intestinal functions, or if Mr. Eliot's shadow boxing with passion, or if the lubricities and sexual neuropathies of our too modern writers could subtly invade the brains where they were engendered with locomotor ataxia or paresis. Indeed, for all I know, perhaps they can. And there is no arsphenamin for the psychic treponema.

Typhus is far less perilous.

CHAPTER III

Leading up to the definition of bacteria and other parasites, and digressing briefly into the question of the origin of life — a discussion without which the reader would be quite unprepared for what is to follow

I

In the history of the immense universe, that of our little planet is an isolated and probably unimportant episode. On some older island in the immeasurable spaces, some other evolution may have produced beings so much wiser than ourselves that they can comprehend the origin of life. For there is no just reason to believe that we — transitional creatures in the upward progress of evolution — have reached the highest possibilities. The tragedy of man is that he has developed an intelligence eager to uncover mysteries, but not strong enough to penetrate them. With minds but slightly evolved beyond those of our animal relations, we are tortured with precocious desires to pose questions which we are sometimes capable of asking, but rarely are able to answer. We have learned to dream of conquests of the forces about us; we investigate matter and the energy that moves it, the order that controls the worlds and the sun and the stars; we train our minds inward upon themselves, and discover emotions, ethical desires, and moral impulses — love, justice, pity

— that have no obvious relation to mere animal existence. The more we discover, the greater is our hopelessness of knowing origins and purposes. The more our ingenuity reveals the orderliness of the nature about us and within us, the greater grows our awe and wonder at the majestic harmony which we can perceive more clearly with each new achievement of art or of science, but which — in ultimate causes or in goal — eludes us. To feel this awe and to wish to fit into the harmony of natural things, with a vision of the whole, is apparently a definite phenomenon of human psychology; it is the force that has engendered religions, just as the instinct to understand the material environment has produced science, and the impulse to express æsthetic reactions has produced art. It is obvious that religion begins where philosophy takes off from the solid shore of the exact sciences into speculative waters, the shallows of which are metaphysics. It is not entirely sensible in modern times, however, to speak of conflicts between religion and science which, to truly civilized people, have not existed for a long time. When perturbed ministers, like the Reverend Dr. Fosdick, passionately deny such a conflict, they are pounding the table and asserting that the earth is round. They desire to preserve the beneficent social and moral influences of an organized church in a world not yet ready for a purely ethical code. And when distinguished minds, like Millikan and others, take wing from the ultimate peaks of exact science into the stratosphere of an old-fashioned heaven, they illustrate the biological truth that the mind of man possesses ethical desires which the most highly developed knowledge of science cannot satisfy — obviously, never will satisfy.

It is not entirely a matter of accident that astronomers, physicists, and mathematicians are more prone than the biologist to fall into the lap of Mother Church or at least into that of one of her barren metaphysical sisters. The biologist, in his work, is always confronted with the mystery of life. He learns a reverence for it which, compounded of wonder and awe, keeps him modest and willing to admit without despair that here is something quite amazing, worthy of continuous study, but, for the time being, beyond his capacities to comprehend. The sagacious physicists to whom I have alluded scamper back to God. But they think they have reached a new understanding and have discovered a new and modern Jehovah, when as a matter of fact all they have done is perhaps to take away his beard and express his thunder in ergs. In their hearts and minds he still remains the same old "Almighty." What might eventually be attained is what, for a time, the Greeks achieved when the philosophy of Plato was the religion of educated people, and what, in the form of Confucianism, existed to some extent in China.

This, however, is too much to hope for in our present overpopulated world, for as fast as ministers like Dr. Fosdick throw overboard their ballast of mysticism in order to cross the shoals into a quiet harbor of reason, Millikan and other physicist-metaphysicians fish it out again to steady them in making the high seas of speculation. The prospect is hopeless unless someone can appear who will be as rigid as was Christ in differentiating between issues of the spiritual and the material, and who at the same time possesses a thorough familiarity with the possibilities and limitations of modern science.

The scientist who achieves intellectual and emotional maturity without losing his investigative vitality and courage — that is, without metaphysical surrender — can come to rest in philosophical tranquillity with the recognition that science, however highly developed, may never answer the ultimate questions; but that there may be happiness in contemplating nature's orderly coördinations, and peace in modest fellowship with the rational and humane spirits who, throughout the brutalities of history, have held to the purpose of reason. Complete comprehension could add very little.

Bergson suggests that on another planet life might have been evolved by systems entirely different from our own. The element characteristic of substances that supply energy might have been other than carbon, and the element characteristic of living matter might have been other than nitrogen, leading to living bodies radically different from our own in chemistry, anatomy, and physiology. This may perhaps be true; but to believe it would require assumptions to which earthly observations give no clue. The origin of life, so far as we can analyze it on earth, is made possible by the unique properties of the combining powers of three elements,[1] and the infinite diversity of the phases and systems made possible by the properties of water. By these relations, says Henderson, "the pathway from the simple compounds of the atmosphere to the complex organic bodies is a direct one."

[1] Lawrence J. Henderson, *The Order of Nature.*

Out of these combinations and dissociations, in contact with the other elements in the infinitely variable conditions of pressure and concentrations, with the radiant energy drawn from the sun, — somewhere, at some time, — life was begotten. In that transition between the dead organic combination and the similar one that is alive lies the great, incomprehensible mystery. What came before we can reasonably trace; what came after is at least open to inquiry in the records of existing living forms. In that leap from the dead to the living lies the mysterious break of continuity which defies our understanding. Between the chemically definable protein molecule and the living bacterial cell there is a gap of understanding far greater than that between the first living cell and man.

It is not easy to define life. An enzyme that could expend energy and build up new energy for that which it expends, in automatically regulated cycles, would be alive — though soluble and not organized in cellular form. There are invisible agents, parasitic upon plants and animals, which we know only by their activities. The ultramicroscopic virus agents, the mosaic disease which infects tobacco and potato plants, those which cause foot-and-mouth disease, rabies, yellow fever, infantile paralysis, smallpox, and many other destructive maladies, thrive in the living cells of higher beings and reproduce themselves in infinite generations, remaining true to type in habits of specific parasitism. Yet they are so small that they do not interfere with the waves of visible light,[2] but are surely

[2] Ultrafiltration measurements give them magnitudes ranging from 20 to 200 millionths of a centimetre.

large enough to contain a hundred or more of the smallest
protein molecules. It is probable that some of the largest
ones have been seen as just visible dots under the highest
magnifications; but many of them have never been seen.
It is assumed that they are living things, cellularly organized, but we are not sure of this; and the thought is
at least reasonable that some of them are transitional
things between true enzymes and formed cell-individuals.
The evolutional transition from the dead organic complex to the cell may well have been a gradual one of
infinitely small steps which may yet be uncovered. Modern
observations of the bacteriophage phenomenon have at
least given us the material for hopeful inquiry.

Did life originate spontaneously by such progressively
complex associations of matter through enzymes — un-
formed, regulated intermediaries, capable of building up
and expending energy? Or did it come to our earth from
elsewhere, — cosmically, — in which case it would have
had to possess the capacity of resisting, without destruction, exposure to temperatures ranging from absolute
zero to incandescence. We cannot deny these possibilities,
but we have no clue to either. We are beginning to know
that all the processes which take place in living beings
are governed — though with more complexity — by the
same physicochemical laws which govern the reactions
in dead chemical systems. Yet this purely mechanistic
understanding is insufficient for the final answer, and
vitalism is reborn again and again to bridge the gap.

With us, in the same modern world in which we cultivate what we call art and science, our almost ultimate

ancestors, the Protozoa and bacteria, have survived. The bacteria particularly (nearest of recognizable cells to the stem of living things) are still more important than we. Omnipresent in infinite varieties, they perform fermentations and putrefactions by which they release the carbon and nitrogen held in the dead bodies of plants and animals which would — without bacteria and yeasts — remain locked up forever in useless combinations, removed forever as further sources of energy and synthesis. Incessantly busy in swamp and field, these minute benefactors release the frozen elements and return them to the common stock, so that they may pass through other cycles as parts of other living bodies. Some of them correct the excessive enthusiasms of their too thorough brethren, which break down nitrogenous substances to free nitrogen. In the soil and in the root tubercles of clover, peas, and other legumes, bacteria are busy fixing nitrogen into complexes ready for revitalization. Without the bacteria to maintain the continuities of the cycles of carbon and nitrogen between plants and animals, all life would eventually cease, plants would have no nitrates and no carbon dioxide with which to grow, cows would have no clover to eat, men would have no beef and vegetables. Without them, the physical world would become a storehouse of well-preserved dead specimens of its past flora and fauna — as useless for the nourishment of the bodies of posterity as ugly and stupid thinking, petrified in books, is useless for the nourishment of its spirit.

2

Among the adages and proverbs which tend to become the philosophy of the thoughtless, one of the most dangerous is: "Seeing is believing." For thousands of years, wise men believed that the earth was flat and that the sun moved around the earth — because they could see with their own eyes that these things were so. It was, in part, this same faith in pure observation which delayed for so many centuries a sensible approach to the problem of the origin of life. Maggots were engendered from decaying horseflesh, lice and fleas from human perspiration; a horsehair in a bucket of water became a threadworm. These things could be observed and, therefore, were true. Even the successful production of the homunculus (ἀνθρωπάριον) was announced by the alchemist Zosimos in 300 A.D. with the same confidence and nearly as much authority as some of our modern biologists announce the transformation of ultra-microscopic viruses into bacteria on similar tenuous evidence.

In spite of the immense literature of error which we shall presently consider, the ancient mediæval speculators were less dangerous to understanding than are their modern representatives. False doctrines became less widely known then, for few people could read and there was little personal gain in notoriety; the public had not begun to become science-conscious and intellectual, and scientific questions were appraised by the intelligent and instructed minority instead of being immediately submitted to the intellectual proletariat. Also, if we feel astonishment at the relatively slight progress that has been made in the solution of the question concerning the origin of life in the course of the many thousand years during which man has pondered it, we must remember that the view of the Greeks in 300 B.C. was a sounder one than any attained until very modern times, when the Greek method of thought was reënforced by the development of biochemical and biophysical methods after a century of a biological clearing of underbrush.

It is interesting to speculate what the Greeks might have achieved in another three or four hundred years of development if the empire building of the Romans, and the evolution of a Christian Europe out of barbarism, had not interrupted them. The one thing the Greeks lacked for the rapid acquisition of the necessary fundamentals of chemistry and physics was an experimental methodology. And this, it would seem, must have inevitably developed out of their geometry — as, indeed, it had already begun to do with Archimedes and a few others. It was the influence of mathematical thought which, in later centuries, gave rise to the method of the experimental isolation of individual phenomena or their fractions. The Greeks were certainly closer to this in 300 B.C. than the Europeans were until 1500 A.D.

The world being as large as it is, it is probably necessary every now and then to mark time culturally for a thousand years or so. And this is what seems to have happened in the single cycle of which we have historical knowledge. The Roman genius for organization and the influence of a supernaturally enforced — and therefore more easily comprehensible — system of Christianity were necessary to bring the hordes of *sans-culottes* of the European forests slowly to the point where, in two thousand years, they might continue where the Greeks left off. As a matter of fact, while European civilization, from 1600 on, went far beyond the Greeks in scientific discovery, it is debatable whether in spiritual and moral development we have yet attained the standards of the Platonic philosophy, which was free from any scaffoldings of doctrine or supernatural buttresses. And in spite of all progress, our school-teachers have substituted "household economy" and "sexual hygiene" for classical history and philology, and the civilized world still continues to support a sort of dole system in the Protestant clergy. Just how badly the cultural spirit of the world has been damaged by the late war, it is too early to say. At the present writing, it certainly looks as though Fascism in Italy, however successful economically, had brought scientific and artistic production almost to a standstill; Russia's science and art have so far been little more than feeble instruments of propaganda; and the present state of the lovely structure of scientific idealism of the Germany of the 1890's brings tears to the eyes.

3

It is significant of our helplessness that the views we hold to-day regarding the origin of life are closer to its revelation only in direct proportion to the refinement of method which science has developed. Our forefathers based their opinions on the testimony of their five senses.

We base our own on the additional reënforcements of chemical analysis, microscopic evidence, the potentiometer, and the thermodynamic laws. In the wake of Pasteur, Darwin, Emil Fischer, Willard Gibbs, and countless others, we are differentiating the problem. One of the great beauties of the scientific occupation is the pride of being a private in the great army of differentiators — the generals of which are never dead to their followers. Every objective gained, every trench dug, every citadel conquered, is a permanent advance in organizing the new territory for the coming of the next integrator. Some day he may arrive and make a dead complex live. He may be the son of an English lord, of a Czechoslovakian peasant, of a Russian Jew, of a French barber, or — most unlikely — of an American broker. Thus is science the great democratic adventure. But when he comes, he will be hailed as King.

The great mystery of life will be revealed as a physicochemical process. But we know already that it is — though we have not succeeded in imitating it. And when we do, we shall be — philosophically — just about where we are now.

Its quest is a sort of forlorn hope of human endeavor, indulged in by the intelligent impractical of every age. But it is a strange fact that the impractical among mankind are remembered. Why? Because of that quality which more than any other lends dignity to life: the instinct for happiness in understanding, — whether it be by intellectual or emotional perception, — which is the most incomprehensible of the attributes of mankind, and which neither the brutalities of individual nor the brutalities of national competition have ever succeeded in annihilating.

Among the impractical quests of man, none has been more alluring than that concerning the origin of life.[8]

In ancient China, insects were produced from wet bamboo in sultry weather.

The ancient Indians (the Laws of Manu) divided the animal kingdom into the egg-born and the "sweat-produced," or flies, beetles, worms, and so forth.

Out of the mud of the Nile, by the heat of the sun, were engendered frogs, toads, snakes, and mice — for could one not see them oozing out of it in the warm months?

The sacred, coprophagous scarabæus was mysteriously fashioned out of balls of dung, and bees sprang from the putrefying cadavers of cattle.

Thales, one of the seven wise men of Greece (an old woman made fun of him because, when he walked out to gaze up at the stars, he fell into a ditch; and his mother kept him from marrying, because when he was young she said, "It is too soon," and when he grew old she said, "There is not time enough left"), thought that water was the source of all living things and that life arose in the warm mud and ooze of the floor of the oceans. He was followed in the same thoughts by Anaximander and Xenophanes. Rain water was added by Anaxagoras, which carried down fertile seeds from the infinite spaces. There seems to have been a general agreement on mud.

That new creatures were born from the union of their

---

[8] An extraordinarily complete and learned compilation of the subject, from which we have freely quoted, has been published by von Lippmann.

similar ancestors was not denied. But, in addition, new ones were being constantly added from the synthesis of sun-warmed organic matter.

Parmenides, Empedocles, and Diogenes of Apollonia favored mud and moist earth as the sources whence life sprang.

Democritus, Epicurus, and their recorder, Lucretius, started something new. Everything on earth has life. The earth is the mother who, in her youth, gave birth to all living things — performing miracles of fecundity which gave origin to plants and animals and even to man. But as she grew old much of her power was lost, and only trivial things like insects, reptiles, and other inferior beings were begotten from decaying organic matter, with the help of warm rain and sunlight.

Plato was reasonably agnostic in these matters, as was Socrates, though the latter invented "Entelechia," the power of the spirit, which, infused into matter, gave it life.

Archelaus believed that the putrefying spinal cords of animals and man were transformed into snakes.

Diodorus, about 30 B.C., revives the old louse story — its origin from human skin and perspiration; and he again asserts that mice were produced from the mud of the Nile, for he could see them slipping out — perfectly formed in front, but unfinished behind.

Vergil seems to have believed the old story about the origin of bees from the dead bodies of steers. It is astonishing, in this connection, that Homer — in the Nineteenth Book of the *Iliad* — lets Achilles speak of the danger of

flies slipping into the open wounds of Patroklos and there
producing maggots — perhaps the earliest exact observation in this matter.[4]

Ovid has the same ideas as Vergil, only he thinks that
wasps come from the dead bodies of horses and beetles
from those of asses.

With the influence of Christianity, there was of course
a considerable change in some of the views. Gregory of
Nyssa, in the fourth century, sticks to the Bible and states
that the beasts and the plants were suddenly born from
the earth by God's will; whereas Augustine was troubled
by his logical mind to the extent of wondering whether,
if the earth retained its power to bring forth animals by
spontaneous generation even after the flood, the Ark
would have been unnecessary; and he could not harmonize
his belief in the goodness of God with the divine production of disagreeable things like mice.

All through the Middle Ages, the same type of reasoning persisted. There was a little less naïveté in some
of the theories, but many others were more fantastic than
anything antiquity was able to produce. The great physician Avicenna believed that intestinal parasites were all
produced from putrefying materials and moisture, and he
completely accepted the origin of animals from properly
combined elements. Lippmann credits him with the statement that, as the result of a thunderclap, an incomplete
calf dropped to earth from the sky.

[4] "But I have grievous fear lest, meantime, on the gashed wounds of
Menoitios' valiant son, flies light and breed worms therein and defile
his corpse — for the life is slain out of him — and so all his flesh shall
rot." (Lang, Leaf, and Myers' translation.)

Even the great Albertus Magnus, in his description *De Animalibus,* adheres to the old ideas that many of the lower animals spring from the materials on and in which they were found, — worms from rotting wood and refuse; bees and beetles from decaying fruits and leaves, — and he seems even to have believed the story about the transformation of a horsehair into a spindle worm — a supposition which is still prevalent among a good many intelligent people. The pious William of Auvergne, Bishop of Paris, was quite willing to believe that worms and frogs were produced in this way, but questioned the matter in connection with horses.

A remarkable tale that kept cropping up again and again until relatively modern times was the belief in the origin of wild ducks and geese from barnacles. These birds came and disappeared and were never seen to breed, so that their origin became the subject of much speculation. One of the stories traced to Saxo Grammaticus was to the effect that the little geese came out of shells which grew on trees in the Orkney Islands. The tale persisted until the latter part of the sixteenth century, when a Dutch sailor penetrated to the Arctic Ocean, where he observed and reported the nesting and breeding of the birds.

Similar to this tale of the barnacle geese is the story of de Mandeville, who, in his *Travels,* speaks of a tree which bore huge, melon-formed fruit of which he himself had eaten, and in which, when it was opened, he discovered a lamb. When the fruit ripens and falls, the lamb's legs become attached to the ground, and it eats all the grass within its range. De Mandeville is now known to have been one of the most talented liars of history. The descriptions of travelers who began to penetrate, in the late Middle Ages and early modern times, into all corners of the earth are responsible for innumerable stories of the same kind. The story of the vegetable lamb was not completely exploded until Linné, in the eighteenth century, examined specimens of the various plants that were supposed to blossom as lambs.

The ideas of Paracelsus were, in regard to the origin of life, not materially different from those of his contemporaries. However, the φύσις of Hippocrates was associated with the Christian belief in the soul in explaining the manner in which God infused life into some of his creatures.

Bacon was a firm believer in spontaneous generation, and Harvey, in 1651, must be regarded as the first who clearly opposed the older views with his famous *Omnia ex Ovo.*

Kepler, wise as he was, believed that plants could grow out of the earth without ancestors, and fish could be produced by spontaneous generation in salt water, just as comets could arise in the skies.[5]

There is practically no attempt through all this period on the part of the most powerful intellects to approach the problem by experimental methods, until the last half of the seventeenth century. In this period, a Tuscan physician, Francesco Redi, published experiments on the

[5] It is to Kepler's credit, however, that — although one of the most eminent physicists of all time — he never wrote a book on God and the Universe.

development of insects, in which he showed that rotting
materials are nothing more than the convenient nest for
the depositing of eggs. He also asserted that various skin
diseases are produced by parasites, and not the other way
round; and Swammerdam comes to the same conclusion
by the convictions of piety, since he held it impossible that
flies, in which there has been expended so much wisdom
and art on the part of Almighty God, could have arisen
by chance from refuse. The honors are with Redi, though
the conclusions are the same.

Leibnitz, in 1714, expresses the conviction that spontaneous generation is impossible, and that neither plants
nor animals could have originated from a chaos of putrefaction. Leibnitz was frankly agnostic in other expressions
on this problem.

Descartes, who was familiar with the work of Leeuwen-
hoek and of all other important naturalists of his time,
gave little thought to the origin of living things, but
speculatively hit the nail on the head by taking for granted
that there may be a world of minute living things from
which life of other kinds can develop by a sort of evolution.

Between the end of the eighteenth century and the beginning of the nineteenth, an accumulation of accurate
observations began to limit the field of speculation, and,
indeed, in surveying the history of the thoughts of men
upon this problem, it is quite apparent that here — as in
all sciences — there has been an inverse ratio between
speculation, on the one hand, and the accumulation of
observations on the other. The discovery of the methods

of reproduction in fungi and mosses in 1729 by the Florentine, Micheli, and Spallanzani's experiments on insects,
led to an increasing conviction that no such thing as
spontaneous generation could take place. Lippmann mentions the amusing fact that one of the important observations on this subject was made in 1804 by a chef in a
Paris kitchen, Appert by name, who preserved foodstuffs
by heating them and putting them into hermetically sealed
pots — an observation which was in line with a similar
one made by Scheele on the preservation of vinegar by
boiling and sealing in vessels. There were throwbacks,
like Needham, but the modern era had begun and the experimental method was soon to take charge of the development of biological thought.

4

With the gradual development of experimental method,
those who were curious about the phenomenon of life
became, by the very precision of their observations, more
modest in regard to speculation. Modern biology was
born when scholars began to concentrate their complete
attention upon the study of the manner in which life
existed, and limited speculation entirely to the construction of trellises along which new experimentation might
grow. The final demonstration, by Pasteur, that alleged
observations of spontaneous generation were attributable
to experimental error marked the ending of biological
mediævalism. But long before this, chemistry, emerging
from alchemy and physics, turning from the firmaments
to the minor affairs of this earth, had started biology on

its modern career. Thus, biology began as it will end — as applied chemistry and physics.

It will be of profit, in maintaining this thesis, to set forth, in the bare bone, the structure of biology as it has come down to our time. The reader of imagination will remember with sympathetic admiration the unnamed multitude of patient toilers, the unknown soldiers of the great struggle toward the truth, who helped to forge the tools for the hands of genius.

Everyone who thinks about these matters can construct a table of significant achievements for himself, and no two will be alike. But since this book is written more for our own amusement than for anyone who may possibly buy it, we set down in chronological order those conquests of understanding which seem to us to have most directly contributed to the modern views of the mechanism of living things. We give them without explanations, since those to whom such matters are unfamiliar may look them up in any up-to-date history of science.

1774. Priestley recognizes that "spoilt" air (spoilt by mice) was made "good" by the presence of green plants. In 1780, Ingenhousz shows that this action was due to the presence of green plants which acted only under the influence of light; in the same year Senebier demonstrates the change to be one from carbon dioxide to oxygen, and in 1804 de Saussure proves the quantitative nature of the conversion.

1784. Lavoisier demonstrates the indestructibility of matter. Quantitative chemistry begins; respiration is recognized as akin to combustion.

1812. Kirchhoff finds that starch can be converted into glucose by dilute sulphuric acid, without the acid itself being changed. This may be regarded as the first clue to the understanding of catalytic processes, leading to Berzelius's conception of a "new force," in which he saw a powerful factor in the explanation of the chemical processes of the living body.

1821. Cuvier lays the foundation of paleontology.

1824. Synthesis of an organic compound (urea) by Wöhler.

1828. Discovery of the mammalian ovum by von Baer. The birth of modern embryology and the first great forward step in this direction since Harvey.

1838–1839. Schleiden demonstrates the cell structure of plants, and Schwann the cell structure of animals.

1838. Cagniard de la Tour proves that fermentation is dependent on yeast cells.

1838. Von Mohl describes protoplasm.

1840. Max Schultze conceives of it as the "physical basis of life."

1842. Mayer suggests the first ideas concerning the conservation of energy, later developed in an orderly manner by von Helmholtz in 1847 (*Abhandlung über die Erhaltung der Kraft*), the eventual consequences of which were the thermodynamic laws.

1842. Birth of biochemistry with Liebig's volume, *Die Thierchemie,* and so forth, on the application of chemical methods to animal tissues; also containing the important conception of animal heat as combustion.

1857. Claude Bernard lays the foundation of modern

physiology, and discovers the production of glycogen by the liver. The beginning of the application of biochemical and physiological methods to the living animal.

1859. Darwin and Wallace advance the ideas of organic evolution, bringing in their train the energetic development of comparative anatomy, embryology, and rational systematology.

1860. Final refutation of the experiments on spontaneous generation by Pasteur.

1861. Recognition of differences in the laws of behavior of the so-called "crystalloids" and matter in particles larger than molecules. The birth of colloidal chemistry by the studies of Graham.

1862. Pasteur defines the dependence of fermentation and putrefaction upon living organisms.

1865. Mendel's work on the crossbreeding of sweet peas. This work, which would probably have materially modified Darwin's original hypotheses, was completely buried in a local scientific journal until 1900, when it was discovered, confirmed, and extended by de Vries and others. It was the foundation of the science of genetics.

1867. Traube's work on semi-permeable membranes.

1877. Discovery of osmosis by Pfeffer.

1880–1900. Development of modern bacteriology and immunology, with the growth of technique for the study of life in its simplest available form.

1885. The correlation of osmotic pressures with their chemical and physical properties of solutions, by Van't Hoff.

1885. Rubner applies quantitative methods to the study of the heat value of food materials.

1887. Beginning of the syntheses of organic matter by Emil Fischer — glucose, fructose, and finally polypeptide, which is one of the higher cleavage products of protein. With the era of Fischer begins the true structural knowledge of the proteins.

1888. Elucidation of the carbon-nitrogen cycle by Hellriegel and Wilfarth.

1889. First discovery of an ultra-virus (mosaic disease of plants), by Beijerinck.

1893. First discovery of ultra-virus causing disease in animals (foot and mouth disease), by Löffler and Frosch.

1900. Beginning of knowledge of the effect of radiant energy (X-ray, ultra-violet) on life processes.

1902. Sutton first pointed out that chromosome segregation furnished the mechanism by which Mendelian laws could be explained.

1904. Discovery of hormones or physiological messengers; internal secretions defined by Bayliss and Starling.

1910. The significant beginning of the application of physicochemical methods to protein and to living tissues; acid base equilibrium; hydrogen ion concentrations; membrane potentials; Donnan's equilibrium; oxidation reduction phenomena; surface phenomena and electrophysics of cells and fluids of living complexes. Those responsible: Sörensen, Loeb, Henderson, Clark, and many others.

1912. Vitamins discovered by Hopkins and Funk.

1915. Discovery of the bacteriophage phenomenon by Twort and d'Herelle, with the suggestion of the possibility that they may be intermediate substances between the enzyme and the formed cell, having the power of reproduction only in the presence of specific living cells, upon which they act. Whether these substances are alive or dead is at present almost an academic question.

1925. Discovery of the relationship between radiant energy and the accessory food factors; the activation of fats to vitamin functions by radiation with ultra-violet light. Based on experiments of Steenboek and of Hess.

1926–1930. The crystallization of enzymes. Sumner prepared urease in the crystalline form in 1926. In 1930 and 1932 respectively Northrop published the crystallization of pepsin and of trypsin.

All this may seem remote from the story of typhus fever; but only to those who are impatient for the sensational events in a turbulent narrative. Without the developments here recorded, we should now know little about the true nature of the subject of our biography.

CHAPTER IV

*On parasitism in general, and on the necessity of consider-
ing the changing nature of infectious diseases in the historical study of epidemics; with a brief consideration of
syphilis as an illustration of this contention. These matters
have direct bearing on our biography, since we must proceed as though we were writing of a man for readers
ignorant of the race of men*

I

NOTHING in the world of living things is permanently
fixed. Evolution is continuous, though its progress is so
slow that the changes it produces can be perceived only
in the determinable relationship of existing forms, and
in their paleontological and embryological histories.
Though the processes which determine evolutionary
change do not appear as simple to-day as they seemed
when the *Origin of Species* was published, it would occur
to no biologist to assume that any living form is permanently stabilized. On purely biological grounds, therefore, it is entirely logical to suppose that infectious diseases are constantly changing, new ones are in the process
of developing, and old ones being modified or disappearing.

Parasitism originated in dim primordial antiquity as a
consequence of habitual contacts between different living
things. It did not develop suddenly, but evolved gradually, as one form adapted itself, step by step, to the en-

vironmental conditions found in or upon another. Parasitism, in its origin, means a breaking down of that opposition which, normally, every living cell complex offers to invasion by another living entity. The simplest illustration of this (for want of a better name, we may call it "vital resistance") is the well-known one of the frogs' eggs. They develop and remain free from invasion in a pond which is swarming with bacteria and Protozoa. A frost kills them overnight, and within a few hours their substances have become culture media for innumerable microörganisms. It is conceivable — and, indeed, could be supported by experimental evidence — that a diminution of this "vital resistance" — which is, in itself, a complex phenomenon — may let down the bars sufficiently to permit invaders to gain a preliminary foothold, even though the host does not succumb to the injury which rendered him susceptible. And once begun, the further evolution of parasitism can proceed in an almost unlimited variety of directions.

Parasitism represents that phase of evolutionary change which lends itself most easily to analysis. There are few parasites which cannot be traced with considerable clearness to some free-living ancestral stock, either still existent or available in fossil form. From this point of view, the study of parasitic adaptation is one of the most important buttresses of evolutionary theory. Each instance represents a miniature system in which the host is the world by which the parasite is moulded. The parasitism which is infectious disease involves the invasion of more or less complex plants or animals by simpler, in most cases, unicellular, beings — like the bacteria, the Protozoa, the Rickettsiæ, and the curious, still undefinable agents of which we speak as "ultramicroscopic" or "filterable" viruses. Though actually complex in function and metabolism, these supposedly simple things display an amazing biologic and chemical flexibility; and since, in them, generations succeed each other with great speed (at least two every hour, under suitable circumstances), the phenomena of infection constitute an accelerated evolution extraordinarily favorable for the observation of adaptive changes. It would be surprising, therefore, if new forms of parasitism — that is, infection — did not constantly arise, and if, among existing forms, modifications in the mutual adjustment of parasites and hosts had not taken place within the centuries of which we have record.

As a matter of fact, the evidence of modern bacteriology lends much likelihood to the view that epidemic diseases are constantly changing; not, perhaps, with sufficient speed to confuse the diagnostic problems of any particular period, but still rapidly enough to encourage the consideration of this factor in the study of epidemic history. To be sure, it has not — so far — been possible in the laboratory to convert a pure saprophyte [1] into an habitual parasite. But it is relatively easy to induce fatal infection with an organism of ordinarily low parasitic powers by reducing the resistance of an individual host. This has been repeatedly done since the time of Pasteur. Moreover, recent advances concerning what is technically spoken of as "bacterial dissociation" have developed simple

---

[1] If the reader does not understand this word, it is too bad.

methods by which a majority of the highly infectious bacteria can be deprived of their virulence and then reversed to their fully pathogenic conditions. Such changes in both directions occur in the bodies of infected animals, can be produced at will in test-tube experiments, and can be correlated with morphological and chemical changes in the bacteria themselves. The subject is one of the most important fields of contemporary investigation, and the results achieved have profoundly modified conceptions of infection. To pursue it further would obviously lead us into technical discussions, more suitable for a textbook of bacteriology. The matter is mentioned in the present connection merely to support our contention that the historical study of infectious disease must, hereafter, take into account the fact that parasitic adaptations are not static, and that extraordinarily slight changes in mutual adjustment between parasite and host may profoundly alter clinical and epidemiological manifestations.

There is a wide range of delicate gradations between saprophytism and parasitism, and the biological and chemical properties along which adaptation changes progress are — to some degree — dependent upon whether an organism that causes disease in man and animals has retained the capacities for life in nature, whether it passes through intermediate hosts, or whether it is so closely adapted to an individual host that it cannot exist apart from him, and perishes when the host dies, unless transmitted to another.

The last condition is the one in which noticeable modifications can be most reasonably expected within the short

period of human records. In such cases, there is an uninterrupted transmission from host to host, the parasite
is never subjected to environments other than those to
which it is most perfectly adapted, and, in consequence,
evolution may progress in a single direction — toward
a more perfect mutual tolerance between invader and
invaded. It is conceivable that, when such parasitism
first begins, the host's reactions are violent, and either
the invader or the host succumbs, according to complex
criteria which vary for individual cases. As adaptation
becomes more perfect, reaction is less energetic, and disease becomes less severe and more chronic; finally, a
stage may be reached in which mutual adjustment is so
nearly perfect that the host may show no signs of injury
whatever. This condition exists, for example, in certain
trypanosome infections of rats, in the spirochætosis and
sarcosporidial infections of mice, and in a large variety of
other conditions of animals and plants. In these, the infected animal shows practically no signs of discomfort
or pathological change in reaction to the parasite. The
principles have been thoroughly discussed by Theobald
Smith. In animal populations, the first impact of a new
virus is upon individuals of all ages. The survival of
some of them is a matter of chance, depending on genetic
differences or the accidental overlapping of immunity
derived from other — possibly related — diseases. The
extinction of many species of animals in past ages is best
explained by freshly introduced parasites. Subsequent im-
pacts are against the very young, and this tends to elim-
inate the weak variants and leads to a population gradually

growing more resistant to that particular form of infectious agent.

In man, a condition which illustrates these principles is syphilis. There is little doubt that when syphilis first appeared in epidemic form, at the beginning of the sixteenth century, it was a far more virulent, acute, and fatal condition than it is now. Uninterrupted transmission from one human being to another, without intervals of extraneous existence in the course of almost five hundred years, has led to gradual mutual tolerance, one of the consequences of which has been an increasing mildness of the disease. If mankind could be kept as thoroughly syphilized in the future as it has been in the past, another thousand years might produce a condition not unlike the present spirochætosis of mice, in which a peritoneal puncture of almost any *bon vivant* would reveal the presence of a treponema pallidum infection of which the host is all but unconscious. Arsphenamin has probably ruined this prospect.[2]

In those forms of parasitism in which the invading organism, in spite of its capacity for infection, has at the same time retained saprophytic properties, it is less easy to determine changes within the periods of historical record. Anthrax and lockjaw — deadly to man and ani-

[2] This might be a loss to civilization: it has often been claimed that since so many brilliant men have had syphilis, much of the world's greatest achievement was evidently formulated in brains stimulated by the cerebral irritation of an early general paresis. We omit reference to specific instances of this among our contemporaries only to avoid, for our publishers, the vulgar embarrassment of libel suits. Modern treatment, and the agilities of expert testimony, render legal proof of such contentions hopelessly difficult.

mals — can, in spore form, be preserved for years in soil without loss of pathogenicity, so that — reinoculated by accident — they can again cause fatal disease. Typhoid and dysentery bacilli, cholera spirilla, the streptococci and staphylococci which cause surgical infections, and many other microörganisms can survive for longer or shorter periods separated from the host; and the circumstances under which this is possible, the length of time of survival, and the alterations which take place in them during such periods, are all of them of the greatest importance to the student of epidemics. Yet even in such infections by half-parasites — if the infection is widely disseminated — the factors discussed above become active, and successive generations tend to develop increased resistance. For human infections, many examples of this could be cited — one of the most illustrative that of tuberculosis, in which the high susceptibility of aboriginal peoples as compared with resistance of the thoroughly tuberculized populations of European origin is a well-known fact.

The idea that we may logically expect modifications in the clinical and epidemiological manifestations of disease within the short period of human history is especially encouraged by study of the so-called "filterable virus" agents. Not an inconsiderable number of the more important epidemic diseases are caused by these mysterious "somethings" — for example, smallpox, chicken pox, measles, mumps, infantile paralysis, encephalitis, yellow fever, dengue fever, rabies, and influenza, to say nothing of a large number of the most important afflictions of the animal kingdom. Here, as in bacterial disease, there is a lively interchange of parasites between man and the animal world. Indeed, since we can neither see these infectious agents nor cultivate them, except in the presence of living tissues, the only opportunity we have of subjecting any of them to systematic study is by finding some animal in which disease can be produced. As a consequence of such study, it has appeared that these agents, even more than bacteria, are of an extraordinary biological plasticity, and can often be modified by simple laboratory manipulation. The transformation of smallpox virus into vaccinia by passage through cattle is far more profound a change than the alteration which differentiates the plague of Athens from smallpox as we know it today. The mere passage of the virus through another species has — in this case — so altered it that it will no longer cause more than a negligible local reaction in man; but, nevertheless, it retains the fundamental biological properties by which it immunizes him. In the same way, the passage of rabies virus through rabbits rapidly increases its virulence for these animals, slightly diminishing it at the same time for monkeys and man. Yellow-fever virus, injected into the brains of mice, ceases to produce typical yellow fever, but causes a form of encephalitis which, thereafter, can be carried in series from mouse to mouse. Carried back to monkeys, even though passed through mosquitoes, it retains its affinity for the nervous system. As a matter of fact, a large number of these viruses, including that of herpes, which causes cold sores, vaccinia virus, and many others, can, by appropriate manipulation, be adapted to what is spoken of as "neurotropism" — that is, so changed that they will selectively invade the nervous system and cause encephalitis.

What we speak of as "new" disease, therefore, need not be conceived as the acquisition — *de novo* — of forms of parasitism that have not previously existed. While this process is probably continuing, it is too gradual and slow to be traceable from an established disease to its ultimate origin. There remain two chief sources of new diseases within historic periods: namely, the modifications of parasitisms already existing in man by gradual adaptative changes in their mutual relations; and the invasion of man by parasites, well established within the animal kingdom, by new contacts with types of animals and insects to which mankind was not previously exposed. That there are many diseases already existing in nature which man has not hitherto acquired only because of lack of opportunity is quite obvious from the recent experience with the psittacosis of birds and a disease of sheep spoken of as "louping ill." In both of these conditions, although isolated human cases had been observed, laboratory association promptly demonstrated an extreme infectiousness to investigators. The Australian X disease — a poliomyelitis-like condition — was probably contracted by man from sheep, and tularæmia — a disease not recognized before 1904, and at present spreading through the United States — is acquired from a number of animal sources.

One of the most interesting phenomena of infectious

parasitism is the interchange of infectious agents between insects and the world of higher animals. This is a large field, which we have no intention of discussing except in so far as it concerns the subject of our biography — typhus fever. Entirely apart from the medical and sanitary aspects of the typhus-fever problem, the circumstances of its transmission are of extraordinary biological interest, because they give us — more than any other disease cycle — the opportunity of studying the evolution of a parasitism which has taken different channels in various parts of the world, adapting itself to the divergent circumstances of local insect and rodent distribution. Typhus fever is one of the Rickettsia diseases which form a closely related group. The minute, bacillus-like organisms which cause these conditions (Rickettsiæ — named after Ricketts, an American who died while investigating typhus in Mexico) are closely related to a number of similar and harmless microörganisms which are habitually found in the bodies of many insects. It is, for this reason, not unlikely that the original parasitism of these organisms was acquired by insects, and from them was passed on to some of the lower animals (rodents) and so to man. These conditions are discussed at some length in a later chapter.

2

When circumstances are such that an infection can saturate almost the entire population of crowded regions, the result is what the Germans call *Durchseuchung*. The accidentally less susceptible survive, and through gener-

ations a gradual alteration of the relationship between
parasite and host becomes established. The more thorough the saturation, the more apparent the results. The
simplest demonstration of such changes is the rapidity
of spread and the virulence of a disease when it is first
introduced into the reservoir of an aboriginal — that is,
entirely susceptible — population. When measles first came
to the Fiji Islands in 1875, as a result of the visit of
the King of the Fijis and his son to Sydney in New South
Wales, it caused the death of 40,000 people in a population of about 150,000. Another example is the terrific
violence of smallpox when first introduced among the
Mexican Indians by a Negro from the ship of Narvaez.
The virulence of tuberculosis for Negroes, Eskimos, and
American Indians living in contact with whites is another
case in point. Any number of illustrations of this kind
might be cited. But even among crowded, thoroughly infected populations, diseases have changed within relatively
short periods. Scarlet fever has become definitely milder
throughout Western Europe, England, and America since
about 1880. The same is true of measles and diphtheria, as
regards both incidence and mortality. The change began
well before modern preventive methods had exerted any
noticeable influence. Perhaps it is not an accident, however,
that, in the case of diphtheria, — in the control of which
modern bacteriological methods have been most effective
since the late nineties, thus creating interference with
normal evolution, — we are just beginning to observe
the return of excessively toxic and deadly cases, reported
in increasing numbers from Central Europe. It is not at

all unlikely that the successful control of an epidemic disease through several generations may interfere with the more permanently effective, though far more cruel, processes by which nature gradually immunizes a race.

Syphilis best exemplifies the alterations which may take place in a disease within a short period, if the population is once thoroughly "saturated." The problems connected with it are so interesting that they seem worth a few paragraphs. Before the last decade of the fifteenth century, there are few reliable records of syphilis in Europe. The subject has been greatly disputed, and many passages — especially in ancient Hindu manuscripts — have been interpreted as signifying that venereal sores similar to those characteristic of syphilis were known in the ancient world. There are, however, forms of non-syphilitic venereal sores, the so-called "soft chancres" or "chancroids," which cannot be distinguished from true syphilis on the basis of extant descriptions; and no physicians whose writings have come down to us from ancient or mediæval literature describe any disease characterized by the sequence of genital sores, followed by skin eruptions and the various secondary and tertiary lesions, which were obvious enough to the physicians of the Renaissance as consecutive stages of one and the same original cause.

Medical historians have cited many observations which they regarded as indicating the ancient existence of syphilis; but most of these, on close scrutiny, turn out to be unconvincing. Talmudic references are not sufficiently precise to permit conclusions, and such allusions as those of Celsus, in the Sixth Book of his *Medicina*, the regulations for prostitutes issued by the Countess of Avignon in 1347, and similar ones, do not constitute reliable evidence. Ozanam quotes two sonnets from a Florentine poet — one entitled "De Matrona," the other "Ad Priapum" — which he accepts as definite proof that syphilis existed in 1480, when the poems were written. Careful translation of these sonnets, with particular scrutiny of the expressions in them which are diagnostically significant, leads to the conclusion that they are merely very nasty poems, with no precise reference to the disease.

It is not, of course, possible to exclude with certainty the ancient existence of a form of syphilis milder than that which swept over Europe in the early sixteenth century, and Haeser — who does not subscribe to the opinion of the American origin — believes that syphilis may have been prevalent to a limited degree and in a less virulent form since ancient times. Sexual immorality was widespread and quite shameless at many periods of antiquity, in Rome, in the Middle Ages, in connection with the great epidemics, and — a strange and common contradiction between idealism and license — during the period of the Crusades. Gonorrhœa undoubtedly was common all over the known world from most ancient times,[3] and was accurately described as the "running sore" in England, and under the names of *clap* and *chaudepisse* in France. There are unmistakable descriptions of chancroids and phagedenic ulcers, which sometimes extended

---

[3] "No stewholder to keep a woman that hath the perilous infirmity of Burning" (Beckit, *Philosophical Transactions*, xxxi, 47, fourteenth century, cited from Haeser).

widely and destroyed the genitalia; and in these diseases
— as now — there were swellings of the inguinal gland
and the bubo. There are few descriptions, however, in
which it is possible to trace the relationship of a venereal
infection to secondary and tertiary consequences in other
parts of the body. This Haeser is inclined to believe
is due to the unwillingness of doctors and patients to
attribute venereal origin to conditions occurring several
weeks after infection and, similarly, he believes that the
later and usually mild manifestations may have been
overlooked, or described in unrecognizable form. There
are a few accounts cited by him which lend weight to his
views. One, taken from Littré, refers to the observations of the French physician de Berry (thirteenth century), who described a condition venereally acquired
which, beginning in the genitalia, spread to the entire
body: "*Nam virga inficitur, et aliquando alterat totum
corpus.*" Another case is that of Nicolas, Bishop of Posen,
who died in 1382, as a result of "*morbus cancri*" on the
genitals, followed by ulcers of the tongue and pharynx.
A similar case is that of King Ladislas of Poland, and of
Wenzel of Bohemia.[4]

It is thus quite impossible to assert with confidence
that syphilis did not exist in pre-Columbian Europe. But
if it did, it must have been relatively rare, and certainly
so much less virulent than the later malady that the epi-

---

[4] *Wan er Faulen pegan
An der stat da sich dy man
Vor Scham ungern sehen lant.*
            — STEYERSCHE REIMCHRONIK (cited from Haeser)

demic of 1500 marked the beginning of a new phase in the parasitism of the treponema pallidum.

The American origin of syphilis forms the basis of a theory that has become widely prevalent, and although it cannot be proved beyond question that America was the source from which the disease reached Europe, it is more than likely that it existed in the Western Hemisphere and that early explorers may have been infected by intercourse with coastal Indians. In this connection, much has been made of lesions on bones found in the graves of the mound builders of Ohio and other regions — notably, New Mexico, Peru, Central America, and Mexico. Professor Herbert U. Williams, who has recently sifted the evidence, with attention both to the antiquity of examined bones and to the trustworthiness of pathological examinations, believes that there is unmistakable evidence of syphilis in many of these lesions.[5] Williams has also reviewed some of the early Spanish literature bearing on the same question. In the *Life of Christopher Columbus*, by his son, Ferdinand, there are included passages from the writings of a hermit of the order of Saint Jerome, — Pane, by name, — written at the time of the second voyage of Columbus. The passage quoted by Williams reads as follows: —

They say that Guagagiona being in the land where he had gone, saw a woman whom he had left on the sea, from whom

[5] It must always be remembered that some of the lesions observed in the Western Hemisphere and attributed to syphilis may have been due to a disease which is more than a cousin, rather a half brother of syphilis — namely, yaws.

he had great pleasure, and immediately he sought to cleanse himself, on account of being plagued with the disease that we call
French; and afterwards he betook himself into Guanara, which
signifies a place by itself, where he recovered from his ulcers.

Oviedo y Valdés says, among other things, that the disease of Buas (probably syphilis) tormented the first Christian settlers in the West Indies, and adds: "Many times
in Italy I did laugh, hearing the Italians say the French
Disease, and the French calling it the Disease of Naples;
and in truth both would have hit on the right name if
they had called it the Disease from the Indies." He also
speaks of a knight, Don Pedro Margarite, who had been
on the second voyage, as suffering from the affliction, and
regards him as probably one of the infectious foci from
which it spread at court. He says that it "was something
new, the physicians did not understand." Similar evidence
comes from Las Casas, Sahagun, and de Isla. From the
manuscript of the last named writer, Williams quotes a
paragraph not represented in the printed editions, —
left out for unknown reasons, — which is of exceptional
importance. "As has been found by very long and well-
proved experience, and as this island was discovered and
found by the Admiral Dom Cristoual Colon at present
holding intercourse and communication with the Indies.
As it is of its very nature contagious, they got it easily:
and presently it was seen in the Armada itself, in a pilot
of Palos who was called Pincon and others whom the
aforesaid malady kept attacking. And as it is a secret
disease never seen . . ." and so forth.

Whether syphilis originated in Europe or came to it

from America will probably never be decided. The theory of American origin, however well-founded in other respects, meets with an almost unanswerable objection in the shortness of the period which elapsed between the return of Columbus and the syphilis epidemic which broke out in Naples in 1495. Moreover, Julien, a French naval surgeon, has recorded that syphilis was more common among the coastal tribes who were in contact with Europeans than among the Indians of the interior, even in the early days of exploration of the Western Hemisphere. It is not at all unlikely that a mild form of syphilis occurred all over the world, including China (according to Dudgeon) and Japan (according to Scheube), long before the fifteenth century. This is the view favored by Haeser, Hirsch, and other learned scholars.

While, thus, there remain legitimate differences of opinion concerning the problem of origin, there is no doubt whatever that syphilis flared up in a sudden, intense, and widespread manner shortly after the time when Charles VIII of France led his army through the South of Italy against Naples. The city was taken by the French in February 1495, and the disease promptly appeared among the troops and the burghers. As the army dispersed, deserters, camp followers, and demobilized soldiers spread the infection far and wide, and, because of the malignancy and disgusting character of the malady, it was the custom of the day to blame it upon the enemy. Thus it was at first known variously as the "French disease" or the "Neapolitan disease." Benvenuto said he had "the French affliction."

The infection as it occurred in Naples was to all intents and purposes a *new disease* in representing a completely altered relationship between parasite and host, with consequent profound changes of symptoms. Something must have happened at that time, apart from war and promiscuity, — both of which had been present to an equal degree many times before, — which converted a relatively benign infection into a highly virulent one. The history of the subsequent fifty years strikingly illustrates the rapidity with which adaptive changes may take place. It is probable that in all parasitisms these alterations of mutual adjustment begin with considerable velocity, the curve flattening out progressively with the increasing number of passages of the parasite through the same species of host.[6]

But when the disease first broke out in Naples in the army of Charles VIII, it possessed a violence that is unobserved in syphilis to-day. According to Scharfenberg, it was a feverless disease characterized by pustular and vesicular eruptions with extensive ulceration. Though the first ulcerations usually appeared on the genitals, this was not always the case. Primary contact infections occurred on many other parts of the skin, and the disease was often transferred from mothers to children in ordinary association. The ulcerations which often resulted from the eruptions covered the body from the head to the

[6] Fantastic theories as to the origin of syphilis were held in early days. Van Helmont, Ozanam tells us, believed that it was started by the intercourse of a man with a mare that had glanders. Linder thought that it started by a similar relationship with a monkey, and Manard thought it came from marriage with a leper.

knees. Crusts formed, and the sick presented so dreadful
an appearance that their companions abandoned them and
even the lepers avoided them. Extensive losses of tissue
in the nose, throat, and mouth followed the skin manifestations, and in the train of these came painful swellings
of the bones, often involving the skull. The disease itself, or secondary infection, caused many deaths. In survivors, emaciation and exhaustion lasted for many years.
Fracastorius says that some of the ulcers traveled, like
those that are called "phagedenic," and extended even
into the bones themselves, where "gummositates" or
gummata as large as eggs developed on the limbs and,
when opened, contained white, sticky mucus.

Within a little more than fifty years, the disease had
already changed. Fracastorius's *De Contagione* was published in 1546, sixteen years after his syphilis poem.[7] His
description of the disease, its methods of transmission and
course, is so complete and precise that we cannot question
the accuracy of his observations concerning the changes
that had taken place between his own time and the epidemic of 1495. The passage in the Second Book of *De
Contagione* reads as follows: —

I use the past tense in describing these symptoms, because
though the contagion is still flourishing to-day, it seems to have

[7] The renowned poem of Fracastorius was written in 1530, and in
it the disease was given its modern name — that, namely, of the shep-
herd Syphilus. The poem was finished in its earlier form in 1525, and
presented to the Sainte-Beuve of his time, Bembo. Within the next five
years it was rewritten and enlarged, and a third book was added, which
deals chiefly with the treatment of syphilis with guaiac. However, in both
the earlier and the later versions, Fracastorius indicates in an allegorical
manner that mercury is the best remedy.

changed its character since those earliest periods of its appearance. I mean that, within the last twenty years or so, fewer pustules began to appear, but more gummata; whereas the contrary had been the case in the earlier years. . . . Moreover, in the course of time, within about six years of the present generation, another great change has taken place. I mean that pustules are now observed in few cases, and hardly any pains — or much less severe — but many gummata.

CHAPTER V

*Being a continuation of Chapter IV, but dealing more
particularly with so-called new diseases and with some
that have disappeared*

I

IT is obvious that when one searches the ancient and
mediæval literature for the existence of maladies in which
differential diagnosis is difficult even to-day, one is likely
to make many mistakes. Accurate descriptions are rare
and, even when details of symptoms and courses are as
accurate as those to be found in Hippocrates, there is a
total lack of the laboratory evidence which is often indispensable for certainty. The problem is particularly
confusing in connection with epidemic infections of the
nervous system, many of which are generally regarded
as new diseases at the present time. We are inclined to
believe that a few only of these conditions are new in
the sense that a virus is involved which had never infected man before. It seems more than likely that in many
cases the diseases are new in that they represent a previously unknown biological relationship between parasite
and host. What we have said in the preceding chapter
about the changes which can be experimentally produced
in some of the filterable virus infections bears upon this
point.

We have no reliable evidence of the existence of infantile paralysis in epidemic form before 1840, and it seems likely that if a disease of such striking characteristics had existed in epidemic form it would have found its way into the seventeenth- and eighteenth-century literature. In regard to encephalitis (*vulgo dictu*, sleeping sickness), it is equally difficult to find reliable evidence of its existence before the eighteenth century. In 1712, Biermer studied an epidemic in Tübingen which was popularly known as "sleeping sickness," because it was accompanied by somnolence and brain symptoms. The "*coma somnolentum*" observed by le Pecque de la Cloture in 1769 was similar and, like the disease of 1917, was associated with influenza. Ozanam mentions a condition of like nature occurring in Germany in the last decade of the eighteenth century, in Lyons in 1800, and in Milan in 1802. After this time no reliable evidence of any disease of this kind can be found until 1917. In that year, synchronous with the first considerable outbreak of influenza, a group of encephalitis cases occurred in Vienna. Soon after that others appeared in France, Great Britain, and Algeria; then during the latter half of 1918 cases were seen in North America, and by May 1919 had been reported from twenty states — the largest number from Illinois, New York, Louisiana, and Tennessee. To all intents and purposes, this was a new disease to our generation, and up to the present time the virus of this form (lethargic encephalitis) has never been successfully transferred to animals. In 1924 a clinically similar and much more severe malady appeared in Japan, and while it differed only in severity from that reported previously, successful transfer of the virus of the Japanese disease to rabbits marked it as a new and different type. During the summer of 1932, an outbreak of encephalitis occurred in Cincinnati and in certain parts of Ohio and Illinois, which cannot at present be classified, but in the summer of 1933 again a similar disease started in the neighborhood of St. Louis, attacking over a thousand people within several months, killing 20 per cent of them. And the virus of this disease, unlike any of the others, could be transferred to mice. It appears, therefore, as though, within the course of less than twenty years, at least three new types of severe virus infections of the central nervous system had appeared among us.

Vaccination has been practised on millions of people since the time of Jenner, and never before the present generation has the practice of vaccination been associated with any kind of nervous disorder. Within the last twenty years, however, a severe type of post-vaccinal encephalitis has occurred in a few regions of the world, and since we know, by experimental manipulation, that vaccinia virus can be made "neurotropic" in animals, it is not impossible, though not yet certain, that in these few cases peculiar circumstances have permitted an invasion of the central nervous system by the vaccinia virus. This condition develops in such a disappearingly small percentage of the vaccinated that it has practically no importance and certainly is not an argument against the practice of vaccination. On the other hand, it appears to be a new disease and for that reason is cited in this place. Indeed, under circumstances which we do not understand, a large number

of the filterable virus infections may create disturbances
in the central nervous system. Thus encephalitis can occur in the train of measles, smallpox, German measles,
and influenza, and the laboratory infections which have
resulted from investigations of the parrot disease, psittacosis, and the disease called "louping ill," have in both instances taken the form of encephalitis-like conditions.

In searching the literature for ancestral forms of infectious diseases of the nervous system, one cannot overlook a
curious chapter of human affliction — namely, that dealing
with the dancing manias spoken of in mediæval accounts
variously as "St. John's dance," "St. Vitus's dance," and
"Tarantism." These strange seizures, though not unheard
of in earlier times, became common during and immediately after the dreadful miseries of the Black Death.
For the most part, the dancing manias present none of
the characteristics which we associate with epidemic infectious diseases of the nervous system. They seem, rather,
like mass hysterias, brought on by terror and despair, in
populations oppressed, famished, and wretched to a degree
almost unimaginable to-day. To the miseries of constant
war, political and social disintegration, there was added
the dreadful affliction of inescapable, mysterious, and
deadly disease. Mankind stood helpless as though trapped
in a world of terror and peril against which there was no
defense. God and the devil were living conceptions to the
men of those days who cowered under afflictions which
they believed imposed by supernatural forces. For those
who broke down under the strain there was no road of
escape except to the inward refuge of mental derange-

ment which, under the circumstances of the times, took the direction of religious fanaticism. In the earlier days of the Black Death mass aberrations became apparent in the sect of the flagellants, who joined in brotherhoods and wandered by thousands from city to city. Later, for a time, it took the form of persecution of the Jews, who were held guilty of the spread of disease. The criminal proceedings instituted against the Jews of Chillon were followed by a degree of barbarism throughout Central Europe that can only be regarded as a part of the mass insanity of which the dancing manias were a manifestation. These manias are, in many respects, analogues of some of the political and economic crowd hysterias which have upset the balance of the civilized world in modern times. In some parts of Europe the World War was followed by famine, disease, and hopelessness not incomparable to the conditions which prevailed in the Middle Ages. For obvious reasons, in the reactions of our own day, economic and political hysterias are substituted for the religious ones of earlier times. Jew baiting alone seems common to both.

Although it is likely that the overwhelming majority of these outbreaks were purely functional nervous derangements, a certain number of them may have represented early traceable beginnings of the group of epidemic infectious diseases of the nervous system, in which we now include infantile paralysis and the various forms of encephalitis.

In 1027, in the German village of Kolbig, there was an outbreak among peasants which began with maniacal

quarreling, dancing, and hilariousness, but went on to
stupor and in many cases to death, and, in the survivors,
left behind permanent tremors, possibly not unlike the
"Parkinsonian syndrome" which follows encephalitis
lethargica. Hecker has given a detailed account of most
of the reliable historical records. In Erfurt, in 1237,
over one hundred children were taken with a dancing
and raving disease which, again, in many cases led to
death and permanent tremors in the survivors. The most
severe dancing mania began in 1374, in the wake of the
Black Death, at first at Aix-la-Chapelle, soon in the
Netherlands, at Liége, Utrecht, Tongres, and Cologne.
Men, women, and children lost all control, joined hands,
and danced in the streets for hours until complete exhaustion caused them to fall to the ground. They shrieked,
saw visions, and called upon God. The movement spread
widely, and undoubtedly the numbers of the truly afflicted
were enhanced by multitudes of the easily excited, in a
manner not unlike that observed in modern camp meetings and evangelistic gatherings. Yet there must have
been a physical disease in many of the cases, because
throughout the accounts there is frequent reference to
abdominal swelling and pain, for which the dancers
bound their bellies with bandages. Many suffered from
nausea, vomiting, and prolonged stupor. The condition
was sufficiently widespread and important to warrant a
long dissertation by Paracelsus, who tried to classify the
malady into three subdivisions by a system not of sufficient modern importance to warrant review.

The tarantism of Italy, supposed by many of its

chroniclers to have been caused by the bite of the tarantula, belongs to the same category. It probably had little relationship to spider bite. The descriptions left behind by Perotte, in the middle of the fifteenth century, and by Matthiolo and Ferdinando in the sixteenth and seventeenth centuries, are quite clear in indicating that many of the cases of tarantism represented a nervous disease of probably infectious origin. Some of them have much resemblance with hydrophobia. Melancholy and depression, followed by maniacal excitement and motor activity, ended in death, or less fatally in semiconsciousness, with alternating laughter and weeping. Ferdinando's descriptions add sleeplessness, swollen abdomens, diarrhœa, vomiting, gradual loss of strength, and jaundice. By the middle of the seventeenth century, the disease as an epidemic menace had practically disappeared. Schenck von Graffenberg, writing in 1643, says that St. Vitus's dance attacked chiefly sedentary people — tailors and artisans. When it came upon them, they rushed about aimlessly, and many dashed out their brains or drowned themselves. In others, renewed attacks followed periods of exhaustion. Many never recovered completely.

Hecker's account, which is the source of most of the facts here cited, includes extensive abstracts of the mediæval literature which indicate that, in the dancing manias, many things were involved. In great part, no doubt, the outbreaks were hysterical reactions of a terror-stricken and wretched population, which had broken down under the stress of almost incredible hardship and danger. But it seems likely that associated with these were nervous diseases of infectious origin which followed the great epidemics of plague, smallpox, and so forth, in the same manner in which neurotropic virus diseases have followed the widespread and severe epidemics which accompanied the last war.

2

Diseases new to the population of any given part of the world in many cases were "new" merely in their territorial extension, as the result of established communication by discovery or conquest. Yellow fever and dengue fever — transmitted to man by the same species of mosquito (*Ædes ægypti*) — may well have existed for ages in the West Indies and the continent of South America. But no reliable account of the former exists in Western medical history until Dutertre described the outbreaks at Guadeloupe and St. Kitts in 1635, and Moseley reported the epidemic on Jamaica in 1655. Since that time, the disease has appeared in many parts of the world — though not all — where the responsible mosquito exists or can survive. With smallpox, as Audouard makes clear, it was probably widely distributed by the slave trade, and, in view of the discovery of yellow-fever foci in West Africa, we shall probably never know whether it came to the Americas from there or vice versa. A serious modern problem is that arising from the automobile and aëroplane traffic now developing across the Sahara between Mediterranean North Africa, where the appropriate mosquitoes are plentiful, but which is not yet infected, and the West African coast, where the fever is firmly established.

As far as dengue fever (breakbone fever) is concerned, there is no information of any corresponding epidemic malady until the last twenty years of the eighteenth century. Then, according to the researches of Hirsch, it appeared in many places in rapid succession: 1779 in Cairo; 1780 in Batavia (reported by Boylon); in the same year in Philadelphia (described by Rush); 1784 in Spain. From 1824 to 1827, the first great epidemics were reported from India and from the West Indies and the Caribbean coast, respectively. Since that time, it has been prevalent, in varying intensity, in most of the tropical and subtropical regions of the world. It is not at all impossible that dengue is not in any sense a new disease of the eighteenth century, but was present much earlier, though unrecognized and wrongly regarded by early Spanish writers as a mild form of yellow fever.

In the so-called "new" disease called tularæmia, we have a problem of a different sort. Can man acquire a novel type of infection, so late in the history of a crowded planet as the twentieth century, by contact with infectious agents long established in insects and wild animals? In 1911 a curious plaguelike infection in ground squirrels was found by McCoy and Chapin. After a great deal of difficulty, they managed to isolate a bacillus roughly similar to the plague bacillus, but still quite easily distinguished from it by appropriate methods. It was not until 1914 that the first proved infection of man was reported. Francis names the disease "tularæmia" because the ground squirrel in which the disease was first observed had come from Tulare County, California. On becoming thoroughly familiar with the symptoms in man, he discovered that cases had been reported in 1907 from Arizona and in 1911 from Utah. Since that time, the disease has been found in every state except Maine, Vermont, and Connecticut. In nature, it is an infection of the ground squirrels of the Rocky Mountain states; of wild rabbits and hares; of wild rats in Los Angeles; wild mice in California; quail, sage hens, and grouse in Minnesota; sheep in Idaho; wild rabbits in Japan, Norway, and Canada; water rats in Russia; sage hens and grouse and wild ducks in California and Montana. Many animals that are not naturally infected are experimentally susceptible. Man acquires the disease by direct contact with the infected animal tissue — especially hunters, butchers, and all who handle, skin, and dress infected animals. The infection passes through small wounds in the skin and may be rubbed into the eye with an infected hand. Almost all investigators of tularæmia have acquired it. Among animals, the disease is transmitted by blood-sucking insects, chiefly ticks and flies. It may be transmitted to man by the horsefly and the bite of the wood tick. In ticks, the disease may be hereditary, so that it is not necessary for a tick to bite an infected animal in order to become dangerous to man. Thus we have another disease of animals which may have caused human infections in small numbers for a long time, and has probably existed in animals for centuries, but which did not become a menace to man until the beginning of the twentieth century.

In the case of the so-called "abortus" type of undulant fever — closely related to Malta fever — it is more than likely that failure of recognition before the present era is due to nothing more than the inevitable diagnostic inaccuracy of former times. Fevers of clinical similarity were known to Hippocrates, and Malta fever itself was described in the early eighteenth century as a diagnostic differentiation of familiar fevers, probably of ancient existence, from similar conditions like malaria and the true enteric fevers. But it was not until very recently (1918) that the similarity of the *Brucella melitensis*, the bacilli which cause abortion in cattle (Bang's bacillus), and a bacillus found in swine was recognized. And it was not until 1922 that bacteriological methods enabled investigators to determine that the milk of infected cattle and the handling of hogs or their fresh meat may produce a disease not unlike that transmitted in the Mediterranean basin with the milk of goats. Since then, these diseases have become public-health problems on our continent and in many parts of Europe. But they are probably new only in the sense that we have been able to "cut out" a new subdivision from an ancient disease group by refined diagnosis.

3

We have seen that the appraisal of the appearance of a so-called "new" disease is fraught with many pitfalls — largely the uncertainty of historic data and the relatively primitive diagnostic methods of earlier days. Nevertheless, even our very superficial discussion of these problems may have supported our thesis that infectious diseases are not static conditions, but depend upon a constantly changing relationship between parasite and invaded species, which is bound to result in modifications both of clinical and of epidemiological manifestations. The principle is illustrated with considerably more precision by a survey of infections which, once widely prevalent, were well described, and which have either become modified or have actually disappeared regionally or altogether. In such instances we possess premises for reasoning of considerable accuracy.

An interesting example of this is the vanishing of bubonic and pneumonic plague from Western Europe.[1] The Black Death, which was mainly bubonic plague, is one of the major calamities of history, not excluding wars, earthquakes, floods, barbarian invasions, the Crusades, and the last war. It is estimated by Hecker that about one quarter of the entire population of Europe was destroyed by the disease — that is, at least 25,000,000. It carried in its wake moral, religious, and political disintegration. This epidemic is an excellent example of the biological phenomena which accompany the process of what the Germans call *Durchseuchung*, which, as we have said, means thorough saturation of a population with an infection. There were, of course, — as we shall mention elsewhere, — formidable plague epidemics in Europe before the fourteenth century, but these — as far as we can tell from the records — did not reach Central and Northern areas within the centuries

---

[1] The history of plague has been ably recorded by many historians. One of the most detailed accounts is that of Sticker.

immediately preceding the Black Death. Resistance to infectious disease, an acquired characteristic, is not hereditary — except in the evolutionary sense of the selective survival of the more resistant. And such increase of resistance by natural selection is not noticeably active, unless the infection continues uninterruptedly throughout centuries and is of such an order that a majority of the infected survive. The Black Death, spreading in Europe, therefore, found an entirely susceptible population, which accounts for its terrific ravages. When its first sweep across the Continent was exhausted for want of victims, it remained endemic, smouldering until relighted by the accumulation of new fuel; and thus it broke out again in 1361, 1371, and 1382. These successive calamities, covering only thirty-four years, illustrate the manner in which an epidemic disease can become progressively less fatal, when it occurs repeatedly in populations that have been thoroughly saturated in immediately preceding years. Statistics are of course incomplete, but the records left behind by Chalin de Vinario, whom we cite from Haeser, are particularly instructive in this regard. In 1348, two thirds of the population were afflicted, and almost all died; in 1361, half the population contracted the disease, and very few survived; in 1371, only one tenth were sick, and many survived; while in 1382, only one twentieth of the population became sick, and almost all of these survived. Had the disease continued, constantly present, and attacking a large proportion of the new generations as they appeared, it might gradually have assumed an endemic, sporadic form, with relatively low mortality. As it is, plague appeared throughout the fifteenth century in Europe, but relatively localized and in incomparably milder form, gradually diminishing until it again broke out in the last European pandemic from 1663 to 1668, reached London in 1664, and was so vividly described by Defoe and — in some of its episodes — by Pepys.

There was an outbreak in Turkey in 1661, which spread first to the coast of Greece and the Greek Islands, then traveled rapidly westward and, more slowly, in an eastward direction. In 1663, it reached Amsterdam, where it killed 10,000 out of a total population of less than 200,000. In the following year it gained velocity, killing about 24,000 in Amsterdam, spread to Brussels and Flanders, and thence to London. In the first week of December, 1664, two Frenchmen died in a house in Drury Lane. No other cases occurred for six weeks. On the twentieth of February, 1665, there was another case; then a pause until April. By the middle of May, the epidemic was in full swing. It is reported by Pepys: —

This day (June 7th, 1665), much against my will, I did in Drury Lane see two or three houses marked with a red cross upon the doors and "Lord have mercy upon us" writ there; which was a sad sight to me, being the first of the kind that, to my remembrance, I ever saw. It put me into an ill conception of myself and my smell, so that I was forced to buy some roll-tobacco to smell and to chaw, which took away my apprehension.

King Charles, rejoicing in the victory over the Dutch fleet, saw more and more houses marked with the terrifying cross, and removed the court from town. Two thirds of the inhabitants fled London, carrying the disease first to other cities along the Thames, and finally throughout England.

The epidemic remained several years in Flanders, passed thence to Westphalia, down the Rhine, into Normandy, Switzerland, and Austria, which it reached in 1668. Throughout the remainder of the seventeenth century, trailers of the disease continued, and lasted well into the eighteenth century. There were localized epidemics in Hungary, Silesia, Prussia, the Baltic Provinces, and Scandinavia. In 1711, 215,000 people died of the disease in Brandenburg; 300,000 in Austria. Another wave spread from Marseilles across Provence in 1720 and 1721. After that, the disease, in severe but localized outbreaks, continued through the second half of the eighteenth century, but was gradually pushed eastward, so that the considerable epidemic which occurred in Russia and the Balkans between 1770 and 1772 failed to make headway in a westerly direction. Russia and the Caucasus continued to suffer up to 1820, but since that time no great plague epidemic has swept beyond Russia, and no widespread outbreaks have occurred anywhere in what is spoken of as the Western World.

This disappearance of epidemic plague from Europe presents one of the unsolved mysteries of epidemiology. The disease has been introduced into various parts of Europe and America again and again during intervening years, but has never shown any tendency to spread in epidemic form. In 1899, isolated cases occurred in Trieste, Hamburg, Glasgow, Marseilles, and Naples — in most cases demonstrably the result of the landing of passengers

and sailors from ships arriving from plague foci. Similar
small group infections have occurred in a number of the
South American harbors. Adding considerably to the
mystery of the situation are such instances as the infections
that occurred in Sydney, Australia, in 1903. In January,
a dock laborer died of plague; and on February 14, dead
rats were found on the quays. Another laborer came down
with plague on the fifteenth of February, after traceable
contact with rats; another on February 26. Within the
next few weeks, the keeper of a hotel close to the harbor
was found to have plague, and by the end of June isolated
cases occurred in the suburbs of the city. Comparable
conditions existed in Melbourne in April of the same
year, with scattered cases. In Adelaide the same thing
happened, and plague-infected rats were found, both in
the suburbs and in the city itself. Still no epidemic occurred. In 1900, the disease was carried to New York,
again without serious results. The existence of plague
among the Chinese in San Francisco was discovered in
1900; and cases in different parts of California, widely
scattered, occurred from then on until the end of the
first decade of the twentieth century. As late as 1907,
twenty-four Chinese of San Francisco came down with
plague, with thirteen deaths, and a few cases were found
in Oakland and Berkeley. In the same way, harbors of
England and the larger cities of Central Europe have
occasionally had plague cases, and plague rats have been
discovered in one of the large European capitals as lately
as 1923. Yet no epidemics have resulted.

The first thought that occurs in explanation is that the population of Europe has acquired considerable resistance. That this is not the case is apparent from the susceptibility of Europeans living in India and other plague centres of the East. We cannot, moreover, attribute the change to any success in the destruction of rats. As for fleas, anyone who has not traveled too luxuriously in Central and Southern Europe during the flea month — September — knows well that there is no dearth of fleas. When all is said and done, we have no satisfactory explanation for the disappearance of plague epidemics from the Western countries, and we must assume that in spite of the infectiousness of the plague bacillus, the plentifulness of rats, their occasional infection with plague, and their invariable infestation with fleas, the evolution of an epidemic requires a delicate adjustment of many conditions which have, fortunately, failed to eventuate in Western Europe and America during the last century. The most reasonable clue lies in the increased domestication of rats. Plague epidemics in man are usually preceded by widespread epizoötics among rats; and under the conditions of housing, food storage, cellar construction, and such, that have gradually developed in civilized countries, rats do not migrate through cities and villages as they formerly did. The exemption of many may be directly dependent upon the greater domestication of rats, which remain contentedly at home, and, as a consequence of this, plague foci among them remain restricted to individual families and colonies.

Closely bound up with the biology of plague is that of leprosy. This disease, well known in ancient times, increased immensely in mediæval Europe. It is assumed that it was widely distributed in Europe by the returning Crusaders, although there are indications that it was present to some extent in France in the sixth century. By the end of the eleventh century, institutions for the segregation of lepers — leprosaria — were common, the first one founded in 1067 in Spain by Ruy Diaz de Bivar, commonly known as El Cid. Under the auspices of the church, similar institutions grew in number and size, so that by the time of Louis VIII, Haeser tells us, there were as many as nineteen leprosaria in the diocese of Troyes alone.

The story of leprosy is a chapter as extensive as that of plague, and would require a volume in itself. The point of interest in our present discussion is that after the middle of the fifteenth century leprosy began to decline, and leprosaria gradually became unnecessary. By the middle of the sixteenth century, only a few centres of the disease remained. In the seventeenth century, it had practically disappeared. Medical histories have attributed this decline to all kinds of vague conceptions, based upon assumptions of improved sanitary conditions, and so forth, but none of these are adequate. The most likely solution of the problem was suggested to us in conversation by Professor Sigerist, who connects the disappearance of leprosy with the immense mortality that occurred at the time of the Black Death and its secondary waves. When the plague struck Europe, with its dreadful destruction of human

life, imm.ense numbers — perhaps the majority — of
lepers had been segregated in institutions, which thus
represented a concentration of relatively susceptible and
weak groups. It is not impossible, as Dr. Sigerist suggests, that most of the lepers of Europe were wiped out
by the plague, and that the few who survived were too
scattered and represented too meagre a spark to revive
the disease. This seems especially likely in view of the
relative noncontagiousness of leprosy, the manner of
transmission of which we do not yet understand, but
about which we know that prolonged and intimate contact
alone gives rise to new cases.

4

The so-called "English sweating sickness" is probably
the most important of those severe plagues that tormented
mankind in brief and terrifying visitations and then completely and inexplicably vanished. The "sweat" came on
with tempestuous speed, and disappeared as suddenly as
it came. There is no mention of a similar fever before
1485 or after 1552.

After the battle of Bosworth, in which Henry VII
gained the ascendancy in England, there broke out in the
ranks of the conquering army a disease that completely
put a stop to the procession of the victorious troops. With
disbanded soldiers, it was carried into London. The speed
of spread can be estimated from the fact that the sickness reached its height in London by September 21, the
battle of Bosworth having been fought on August 22. It
spread over England rapidly from east to west, carried

far and wide by the men scattered from the army. In London it killed, within the first week, two Lord Mayors and six Aldermen. It attacked the young and robust, this being one of the points in which it was similar to the Picardy Sweat, of which we shall have something to say presently. The mortality of this English sweating sickness was such, according to Holinshed, that "scarce one amongst an hundred that sickened did escape with life; for all in maner as soone as the sweate tooke them or in a short time after yeelded up the ghost." The Coronation of Henry was postponed. In Oxford, where Thomas Linacre — who later founded the College of Physicians — was then a student, it was so severe that professors and students fled the University, which was closed for six weeks. This first outbreak remained entirely in England, not even spreading to Scotland or Ireland.

The symptoms of the disease have been described by many writers, and, though minor differences occur, the accounts are in the main consistent. Particularly important is the description by John Kaye, whose famous pamphlet on *The Sweate* was published in 1552. The disease began without warning, usually at night or toward morning, with a chill and with tremors. Soon there was fever, and profound weakness. Accompanying this were cardiac pain and palpitation, in some cases vomiting, severe headache, and stupor, but rarely delirium. Although some writers make no mention of a rash, there are nevertheless descriptions which do so — especially that of Tyengius, whose accounts come to us from Forest, and who relates that, after the perspiration was over, there appeared on the limbs small vesicles "which were not confluent but rendered the skin uneven." The profuse sweating, which was the most noticeable characteristic, began soon after the onset of the fever. Death came with astonishing speed. It is stated that many cases died within a day, and some even within a few hours. A single attack did not immunize, since a number of people had two or three attacks in brief succession.

After a short and violent career, the disease completely disappeared, and we find no mention of it from 1486 until 1507.

The second epidemic was apparently much like the first, but there is not much reliable information available. It again started in the summer — this time in London — and, as Senf suggests, it is not improbable that it may have remained endemic in that city during the inter-epidemic quiescence.

In 1518, the disease appeared for the third time, and with enhanced severity. Again it spread over England, again sparing Scotland and Ireland. But this time it reached the Continent, advancing only to Calais, where — strangely enough — none but the English inhabitants are said to have contracted it. Again it killed many patients within two or three hours, and it brought death to many important men in Oxford and Cambridge; in some towns from a third to a half of the population was wiped out.[2]

The sweating sickness seems to have gained energy be-

[2] It is stated that in some places 80 to 90 per cent of the population died.

tween epidemics, for the most severe outbreak was that
of 1529.[3] This started in May, again in London, and the
terror it inspired was so great that society was disorganized, agriculture stopped, and famine resulted. The disease
swept across the sea to the Continent, where it was first
reported in Hamburg, which it reached in July, probably
with a ship returning from England. In the same month
it spread across Eastern Germany to Lübeck and Bremen;
by August, it had reached Mecklenburg; in September
it came to Königsberg and Danzig; thence it traveled
southeastward to Göttingen, where the mortality was so
great that five to eight corpses had to be put into a single
grave. A curious fact noted by many who described it at
this time is the lateness with which the disease reached
the Netherlands, — that is, four weeks later than its appearance in Hamburg, — although active communication
by sea was carried on equally between both places and
England. In Marburg, the epidemic interrupted the Coun-
cil of the Reformation. In Augsburg, 15,000 fell sick in
the first five days. It reached Vienna during the siege of
the city by the Sultan Soliman and, probably ravaging the
Turkish army, may have had some effect on the raising of
the siege. A little later, it entered Switzerland; but it
never crossed into France.

The fifth and last epidemic of the sweat occurred in
1551. Again it started in England, this time in Shrews-
bury, in April, where 900 died in a few days. It spread

[3] We are using the dates given by Haeser. Those of Hecker, and
many others, differ by one year, owing to the discrepancies between the
English and the Roman calendar.

over the whole county, carried about — as Haeser puts it
— "in the drift of poisonous clouds of fog." A strange observation made at this time, which corresponds to the
previous limitation of the sickness to the English inhabitants of Calais, is the apparent exemption of foreign-
ers in England. Yet the fifth epidemic seemed to follow
Englishmen into other countries, so that many died in
France and the Netherlands. This outbreak of 1551 is
the one that John Kaye described in his famous pamphlet.

Only once after this date (we take our information
from Senf) has a sickness resembling the English sweat
occurred, unless we identify the disease — as many have
done — with the Picardy Sweat. About two hundred
and fifty years after the fifth epidemic, that is, in 1802,
at Rottingen in Franconia, a similar but regionally limited
malady appeared.

It is impossible to identify the sweating sickness with any
epidemic disease now prevalent. Purely on the basis of
synchronous occurrence, Schnurrer and others believe
that it was a modified form of typhus, and it is true — as
Senf points out — that it did not spread into any of the
countries where typhus was prevalent at the time. However, this opinion is not convincing. The sickness remains
an entirely individual condition which could not — were
it to reappear at present — be properly classified with any
of the known infectious diseases. The suddenness of onset,
the rapid death, were more violent than any of the diseases of our day, with the exception of occasional cases of
meningitis or infantile paralysis. While speed and manner of spread remind us of influenza, the apparent absence

of prominent catarrhal symptoms, the lack of a secondary pneumonia fatality, and the nonexistence of successive waves within a short period suffice to separate it from influenza, as it now occurs. Its general characteristics would incline us to regard it as caused by a filterable virus of a variety at present unknown. It is a reasonable speculation that the sweat was due to a virus that had for centuries been prevalent on the Continent in milder form, and in England spread in an entirely susceptible community. This is the only basis on which one can hope to explain the reiterated observation that it was peculiar to the English people, even when they were living in foreign parts. Knowing what we do about the wide general distribution throughout modern populations of the virus of infantile paralysis, with which a large proportion of the population has probably been infected without manifest disease before adult life, it is not fantastic to assume that virus infections may eventually become so widely distributed that, in time, entire populations become immunized; and, eventually, a disease which at first was epidemic and severe may become endemic, modified, milder, and — finally — extinct. This sort of thing is certainly going on in diseases like measles, infantile paralysis, and influenza, which — endemic with us — cause destructive and violent epidemics among primitive peoples when carried among them.

Another disease which seems to have come suddenly out of the blue and which, within less than two hundred years, has almost completely disappeared is the so-called "Suette des Picards." There is some confusion regarding the relationship of this disease to the English sweating sickness, and to the so-called "military fevers." Under the latter term, there were probably included a great many of the well-known eruptive fevers, such as measles, scarlet fever, chicken pox, and so forth. It is impossible to review the voluminous controversial literature dealing with these problems, but there are accurate records which show that a peculiar malady quite unlike any of the now prevalent exanthemata suddenly appeared in Normandy in 1718 and spread within a few years into Poitou, Burgundy, and other regions of Northern France. Opinions of the leading medical historians (Hirsch, Haeser, and Ozanam) are at variance concerning the existence of a similar disease in other parts of Europe before 1718. Haeser believes that foci existed before this date in Alsace and in Turin. But the descriptions of such outbreaks lack precision until 1718. Most students agree that, apart from localization, the Picardy Sweat can be differentiated from the English sweating sickness largely on the basis of the eruption and of the violent mental symptoms accompanying the Picardy disease.

Several excellent descriptions regarding its manifestations in different places and many years apart establish its character as a definite clinical entity. The first of these is the precise account by Dr. Belot of the outbreak of 1718. And this corresponds almost exactly with that of Dr. Vandermonde, who reported the epidemic at Guise in 1759.

The onset was sudden, often with a chill, abdominal pain, and difficulty in breathing. There followed a severe headache, high fever and insomnia, and often great excitement. Profuse sweating began within twelve to twenty-four hours, usually accompanied by violent itching. A rash, variously described as resembling measles (*rougeole*) or erysipelas (which probably means an even reddening something like the rash of scarlet fever), was noticed within the first forty-eight hours. Nosebleeds were frequent and violent. In fatal cases there was delirium, and often death was accompanied by convulsions. Many cases died within one or two days.

After 1718, many local epidemics occurred in France — at first at short intervals, later less frequently — up to the middle of the nineteenth century. In the latter part of this period there were similar outbreaks in Northern Italy and in Southern Germany. Altogether, according to Hirsch, 194 epidemics occurred in France between 1718 and 1804. Nothing is known of the mode of transmission, of the causes which led to outbreaks, or of the reasons for their decline. Boyer, writing in 1751, declared that the disease was not contagious, — that is, there was no evidence of transmission from one case to another, — and in this opinion most observers agree.

Unlike almost all other diseases of equal violence, the Picardy Sweat was always closely circumscribed in the individual epidemics. Most of the outbreaks remained limited to individual villages or towns. In only a few instances did they extend beyond defined localities, though on one or two occasions widely separated districts of France were invaded. Individual epidemics rarely lasted more than a few months.

It is impossible to form any trustworthy opinion concerning the nature of this disease. It does not fit into any of the categories of modern classification. While in some respects it resembles rapidly fatal scarlet fever, the absence of any evidence of severe throat infection renders its identification as this improbable. It was surely not measles or smallpox. The only infection of which the fatal and most violent cases of Picardy Sweat remind us is the fulminating meningococcus infections which are occasionally seen during meningitis epidemics. In such infections — many of which were seen in camps during the late war — the sudden onset, profuse rash, sweating, high fever, and rapid death, often with delirium and convulsions, present a clinical picture closely resembling descriptions of the severest cases of Picardy Sweat. Other similarities between the two are the lack of traceable relationship between cases (masked contagiousness) and the limitations of spread. However, the milder cases — which were apparently in the majority — have little resemblance to meningococcus infections. We can only conclude that we are here dealing with a disease which is either unique or which represents a now unknown form of a surviving disease, modified in the course of time. Typhus can be excluded with confidence because of the sudden onset with shaking chills and the rapidity with which the rash developed (one to two days). The violent itching so frequently noticed is also uncharacteristic of typhus. Moreover, the first Picardy epidemic occurred at a time when typhus in its present form had been well known for several centuries.

A few isolated cases of a condition resembling the Picardy Sweat are occasionally reported by French physicians at the present time, but even if these are genuine, no outbreak — even of limited extent — has occurred since the seventies of the last century.

CHAPTER VI

Diseases of the ancient world: a consideration of the epidemic diseases which afflicted the ancient world, with attempts at making diagnoses which, if they are difficult to make at the distance of a thousand years, are equally difficult under the circumstances to disprove. Though this may appear another unnecessary postponement of our biography, it represents our effort to determine the antiquity of typhus fever

I

THAT bacterial diseases have attacked the higher forms of life since the very beginning is unquestionable.

There are, in the Vienna Museum, remains of prehistoric bears which show unmistakable signs of large abscesses of the teeth and jaws. Reasoner has collected from the paleontological literature a number of descriptions of conditions of bacterial origin occurring in prehistoric animal remains. He mentions the remains of a reptile, Dimetrodon, of the Permian Age (21,000,000 years ago), described by Gilmore, in which there was evidence of chronic osteomyelitis of the spine; also a Jurassic crocodile (14,000,000 years ago), described by Auer, which presented signs of infection in the pelvis, with metastases in the femur, the sacral vertebræ, and the palate. Signs of carious teeth, of possibly rheumatic swellings of the joints, have been found in numerous fossils by Renault, Moody, and others. Evidences of bone necrosis and subsequent hyperostoses are not uncommon in fossil remains.

As far as primitive man is concerned, not much is known — although Raymond described a case of spondylitis deformans and one of arthritis of the knee in neolithic bones of France. There is, however, much doubt as to the antiquity of some of these fossils. The meagre paleontological literature of man furnishes little direct information on this problem. There is, however, a good deal of evidence that bacteria became capable of producing infections millions of years ago, and there is no reason to doubt that man from the very beginning suffered from infectious disease; and at the time when mankind had reached the period of the earliest historical records, infectious diseases of many varieties already existed. And though diagnosis is often difficult, it is certain that epidemics were prevalent thousands of years before Christ.

The diagnostic determination of the various infectious diseases from ancient medical literature presents many difficulties because of the uncertainties involved in determining the meanings of descriptive words, unless these occur many times in different connections. Thus it is often impossible to gain any accurate impression of the nature of a skin eruption, since it is often difficult to know whether the word used should be properly translated as referring to raised surfaces, vesicles, pustules, or ulcers.

In Chinese literature, there is very little descriptive material accessible to the Western student from which opinions can be formed regarding the nature of the prevalent diseases. It is not impossible that smallpox and

some of the exanthemata originated in China and reached
Europe across Persia and North Africa. However, opinions concerning this, as expressed by Wise and by Moore,
rest upon very slim evidence. Moore, taking his information from the oldest available Chinese medical treatises,
believes that smallpox was prevalent in China at the
time of the Tsche-u dynasty, — a period between 1122 B.C.
and 249 B.C., — and Smith, in an article in the *Medical
Times and Gazette* for 1871, cites evidence that the disease occurred during the dynasty of Han, about 200 B.C.,
and was imported from India.[1]

In the ancient Indian writings, the Ayur-Veda (date
uncertain, but surely before 200 B.C., perhaps parts of it
as old as 900 B.C.), and the writings of Susruta, there are
accounts that may refer to tetanus and chorea. Fevers of
various kinds were known — some of them quite surely
malaria, some possibly inflammatory rheumatism and perhaps leprosy, known as "Kushta." An intestinal disease,
interpreted with reasonable accuracy as cholera, was well
known. Haeser, who studied the translations of Wise, finds
evidence also of catarrhal jaundice, of gonorrhœa, and
possibly of tuberculosis. It is of particular interest that
in Susruta's writings there are descriptions of genital ul-

[1] This information is largely taken from Hirsch. The origin of
smallpox, however, is a much disputed problem, which has been a subject of learned dissertations by Krause, Hahn, Werlhof, and many others.
Haeser questions the validity of the evidence advanced for the existence
of smallpox in ancient India and China, though he admits the possibility.
He does not accept, as indicating smallpox, many of the descriptions so
interpreted from the writings of Hippocrates. Unmistakably accurate
descriptions of the disease are found in writings dating from and after
40 A.D.

cerations which Haeser thinks may have been syphilitic.

About ancient Egyptian diseases, we have a good deal of information from the *Papyrus Ebers*, which was written during the reign of King Re-Ser-Ka, approximately 1700 years before Christ. The infectious diseases mentioned were an erysipelas-like condition called "Hmaou," which was treated largely with the feces of donkeys; intestinal worms, and varieties of ophthalmia. Examinations made upon mummies by Sir Marc Ruffer, Dr. Eliot Smith, and Dr. Wood Jones revealed evidences of Pott's disease, and in a mummy of the twentieth dynasty (about 1200 B.C.) there are spots on the skin which might have been smallpox. A similar eruption was found on the body and face of Rameses II. On Rameses V there was a triangular ulcer above Poupart's ligament in the region of the inguinal glands, which might have been a plague bubo or a venereal sore (the disease of kings). In some of the older mummies, in which the abdominal viscera had not been removed, Ruffer observed large spleens which may indicate malaria.[2]

The diseases mentioned in the Old Testament are summarized by Garrison in his *History of Medicine* as including gonorrhœa, leprosy, or possibly psoriasis; in Samuel, enlarged inguinal glands are noted, indicating the probability of plague. In the Talmud, there is mention of conditions of the lung that might reasonably be regarded as tuberculosis; of an abscess of the kidney, and of infections of the female genital organs.

[2] For references to many of these observations, we are indebted to an interesting essay by Colonel Reasoner of the United States Army.

Jehovah seems to have been pretty hard on the poor Philistines. In I Samuel IV, there is an account of a battle in which the Philistines overcame the Jews, slaying about 30,000 of them in what appears to have been a perfectly fair fight. The victory of the Philistines was facilitated by the fact that the Hebrew army ran away, and tried to hide in their tents. The conquerors then took the ark of God (I Samuel v) into the house of their own god, whose name was Dagon, and who was a sort of half fish, and consequently more or less helpless. The Hebrew God then smote Dagon, cutting off his hands and throwing him off his pedestal, so that his face was on the ground. This threw a terrible scare into the Philistines of Ashdod, so that they sent the ark to Gath. Thereupon, "the head of the Lord was against the city with a very great destruction: and He smote the men of the city, both small and great, and they had emerods in their secret parts," and "the hand of God was very heavy there. And the men that died not were smitten with the emerods." This is the sort of thing, of course, which — throughout the ages — has led to what in modern terms we may speak of as "Nazi movements." But the Lord only knows what an "emerod" was. Literally, it is a hemorrhoid — the etymological relationship of these two unpleasant words being obvious; but it is hardly likely that even the Philistines could have had a fatal epidemic of hemorrhoids. The words translated as "emerods" are "ophalim" and "teharim," which mean swellings, or rounded eminences. According to our learned informant, the translation "emerods" depends on a comparison with Psalms LXXVIII. 66, where God is

said to have smitten his enemies "in the hinder parts."
This identification is very early, from Talmudic sources
and in Aramaic translations. "Ophalim," according to
other translators, merely means an elevated, rounded
place. Hastings, in his *Dictionary of the Bible*, does not
believe that "emerods" were hemorrhoids, and connects
this description with bubonic plague. Granting, therefore,
that these words refer to swellings in the private parts,
the controversy merely turns upon whether it was the
hinder end or the front end which was affected. Although the material available is insufficient for diagnostic
accuracy, rounded swellings in these regions, associated
with epidemic spread and high mortality, are suspicious
of plague.[8]

In the time of David, as a punishment for the forbidden
census, there was a severe pestilence, which destroyed
70,000 by sudden death. Most of these people are supposed to have died in one day. No clue whatever to the
nature of this malady is available.

Among the plagues of the ancient Hebrews mentioned
by Josephus, there are none that are described with sufficient detail to justify even an intelligent diagnostic
guess. Of the afflictions visited upon the Egyptians, one
had to do with polluted water, which gave them great
pains; in another an innumerable quantity of lice arose
out of their bodies (since many of them died, a louse-
borne disease like typhus may be suspected, though in
view of the absence of historical data concerning typhus

[8] Preuss, *Medizin im Talmud*, is the foremost authority on diseases
of Biblical times.

elsewhere at this period, this is most unlikely); still another was a fatal epidemic of boils.

There is repeated evidence in Biblical history that the fair competition of other nations with the Jews was always rendered a triumph for the Hebrews by the interference of what, to the others, must have seemed a biased and relentless God. We wonder whether this does not lend a great deal of justice to the opinion of Houston Stewart Chamberlain, who explains anti-Semitism entirely on the basis of a clash between religions. Jewish teachings were widely spread in the ancient world, and if the atrocious vengeance of God on all who opposed the Jews — who apparently were no lilies in their relations with others — were believed, hatred and resentment would be easily understood.

2

Interpretation of the infectious diseases that occurred before the time of the Greeks is, in most instances, largely guesswork. From the Greeks, however, a great deal of accurate description has come down to us, which permits us to form intelligent opinions concerning the symptoms, clinical pictures, and often the epidemiology of the conditions that occurred among them. Although there is much medical information before Hippocrates, it has only occasional bearing on the epidemic diseases in which we are interested. Asclepius, a Thessalian king, son of Apollo, was largely a mythical figure, but that a certain amount of knowledge of infection was prevalent among his later followers is apparent from the isolated places in which his temples were built, and from the laws which — in Delos, for instance — prohibited the burying of dead bodies near the temple. Democritus mentions diseases that were probably epidemic, and Empedocles is supposed to have arrested — by the closure of a crevice in a mountain — miasmas that came from a river. Democritus believed that the epidemic diseases which ravaged mankind were due to the destruction of heavenly bodies, the cinders of which dropped upon the earth. Alcmæon stopped a plague in Athens by the lighting of fires. There is, however, no material for ancient diagnostic opinion, even among the Greeks, until the time of Hippocrates.

Hippocrates was probably not the first great physician of antiquity. Indeed, it is likely that many skillful and sagacious medical men practised in ancient Egypt, where — Herodotus tells us — physicians were even more highly specialized than they are to-day, since often they limited themselves to a single organ of the body. There were dentists, as well as internists and surgeons. Hippocrates, however, is the first great physician from whom we have records and writings which show an approach to medical problems entirely analogous to our own. Indeed, his descriptions of cases in the *Epidemion* are so precise that diagnoses more accurate than the ones he made himself can be deduced from his clinical histories.

The Greeks suffered from a great variety of infectious diseases. Being an outdoor people, living in a good climate, with — at first — no formidable concentrations of population, the earlier outbreaks of contagious disease among them were not of sufficient extent to be noticed by histo-

rians. The medical reader is struck by the absence of any
serious descriptions of epidemics among the Greek armies
of Homeric times, during the early struggles between the
Spartans and the Athenians, and in the Persian wars. The
armies were large, often rapidly mobilized, and they must
have had disease; but neither Herodotus nor others who
deal with this period speak anywhere of the kind of widespread epidemic mortality which one is justified in ex-
pecting. This is possibly due to the fact that any such occurrences would then have been interpreted as the wrath
of enraged deities, rather than as visitations of transmissible disease.

In the time about which Hippocrates writes, we find
mention of epidemics of inflamed eyes at Thasos — very
likely pink eye. There were diarrhœas, with fever and
tenesmus, watery stools, vomiting, and sweating — not
improbably forms of bacillary dysentery. The continued
fevers that occurred chiefly in the autumn and early winter
were, in part, quite clearly due to malaria of the quartan,
double tertian, and æstivo-autumnal varieties. There were
prolonged fevers lasting twenty-four or more days, with
— occasionally — late, nonsuppurating swellings of the
parotid glands, which we can reasonably interpret as
typhoid fever; others which, in view of their interrupted
nature and the cult of the goat in ancient Greece, might
well have been Malta fever. There is one description
which unquestionably refers to an epidemic of mumps —
a mild fever, without mortality, and with bilateral parotid
swelling, dry cough, and occasional swellings of the testi-
cles. Sore throats, with coughs, fever, and often with de-

lirium, may have been either scarlet fever or diphtheria.

In the *Epidemion*, there are a considerable number of case histories, quite as thoroughly recorded, from day to day, as many of our modern ones, upon which diagnostic judgment can be based. In many instances, the observations of Hippocrates are so precise that we can often supply, from modern knowledge, the exact type of infection — not infrequently the microörganism that must have been responsible for the individual conditions. In regard to many nonsurgical conditions Hippocrates did quite as well, we surmise, as will be possible for the modern general practitioner or "family medical adviser" who is so dear to the hearts of many of our reactionary contemporaries, and who, by a return to medical muzzle-loading, is to emancipate our profession from all the newfangled laboratory doodads.[4]

Herophontos came down with an acute fever, with liquid and bile-colored movements, tenesmus, and abdominal tenderness. On the fifth day, he became delirious and began to sweat, with continued liquid movements. On the ninth day, there was a crisis with severe perspiration, and a relapse seven days later. Herophontos must have had either acute bacillary dysentery, typhoid or paratyphoid fever, or cholera; but, since his was an isolated case, it was probably not cholera.

The hæmolytic streptococci were as formidable then as they are now. The wife of Philinus and the wife of Domadeos unquestionably died of what we should now call puerperal sepsis.

[4] See Frothingham.

The wife of Epicrates developed a sore throat two days before childbirth, had a prolonged fever, which lasted without abatement for twenty-one days, and did not completely subside for eighty days. She might have had typhoid fever, or a subacute streptococcus infection.

Criton, of Thasos, had a sudden pain in his big toe, followed by fever and delirium on the same night. The next day, his foot was red and œdematous, with little black spots, and his leg began to swell. He was dead in two days, and without doubt died of a virulent streptococcus infection, perhaps starting from an ingrown toenail.

A Clasomenian had what was unquestionably typhoid fever.

A pregnant woman, three months with child (the thirteenth case in the First Book), suffered from a sudden pain in the back, rapidly followed by fever, headache, pain in the neck and right hand, and loss of speech. There was delirium on the fifth day, and paralysis of the right hand and arm. There is no statement as to residual paralysis after recovery on the fourteenth day, but the whole story sounds like acute anterior poliomyelitis, or possibly the encephalitis lethargica which we have thought to be a new disease.

An unnamed man died of a condition which was without much question an attack either of acute appendicitis or of cholecystitis. In the middle of the night, after a heavy meal, he was seized with sudden vomiting, fever, and pain in the right hypochondrium. The symptoms continued; the abdominal pain became general, and he died on the eleventh day. We favor acute appendicitis, because of the omission of any reference to jaundice. It is interesting to note the care with which physical examinations were made by Hippocrates. He states that on first observing this patient, he found no abdominal rigidity. This must have developed later, or we must assume that even Hippocrates may have made a mistake.

Among the remaining cases there are carbuncles, erysipelas, possible diphtheria, various forms of paralysis, and, not impossibly, cases of plague,[5] since there are

[5] If the cases described by Hippocrates were true plague, it is of course strange that there is no description of epidemic spread. That he knew plague in isolated cases seems likely from passages in his *Aphorisms*, cited by Littré, in which he says that fevers with buboes are all dangerous except those which last a very short time. The same author also cites a sentence from the Second Book of the *Epidemion*, which indicates a knowledge of true plague. Hippocrates was born at Cos, in the first year of the eighteenth Olympiad — that is, 460 B.C. The great plague of Athens occurred in 430 B.C., and if this had been an epidemic of bubonic plague, Hippocrates would have recognized it as such. As we shall see in another place, notwithstanding the opinion of Ozanam and some others, the Athenian plague cannot, in the light of the descriptions, be regarded as plague. There was also, during the lifetime of Hippocrates, a severe contagious disease in Persia. Artaxerxes sent envoys to the great physician, offering him rich treasure if he would come to the aid of the stricken Persians. Although (it is so stated, but also contradicted) Hippocrates declined this mission from motives of patriotism, the nature of the Persian disease must have been thoroughly described to him. It is likely, therefore, that if plague in its typical manifestations had existed in Greece during the fifth century B.C., Hippocrates would have described it recognizably. The question has been thoroughly sifted by all the leading medical historians. If Greece was exempt from epidemics of plague at a time when it was prevalent elsewhere, this may have been due to the scarcity or possible absence of domesticated rats. In our chapter on the history of the rat, we discuss the information on which this surmise rests. However, there may have been other, more mysterious reasons. We are faced with a similar problem in the absence of epidemic plague from modern England and

descriptions of buboes of the thighs.[6] There were pneu-
monias and pleurisy, and protracted diseases of the lungs
which resemble pulmonary tuberculosis. Rheumatic fever
does not seem to have been unknown, but the descriptions
are vague.

Our primary purpose in examining the clinical histories
of Hippocrates was to find evidence of the early existence
of typhus fever. Ozanam and others have stated that Hippocrates described typhus fever, and the case that has
often been cited as evidence for such an assumption is that
of the second patient in the First Book of *Epidemion.*
This individual, Silenus, "son of Eualcides, who lived near
the platform, was attacked by a fever as the result of
fatigue and excessive drinking and exercise. From the
beginning, he had pain in the back, headache and pain in
the neck." For a number of days he had fever, with
intestinal symptoms, feelings of pressure in the abdomen,
insomnia, and delirium — all of which might be consistent with a number of different types of infectious
disease, but are quite consistent with the onset of typhus.
On the seventh and eighth days, he had severe sweats,
and on the eighth day he developed an eruption of red,
spherical spots which continued without suppuration. He

Western Europe. Isolated cases of plague have been observed in some of
the larger European cities within the last twenty-five years, but not even
local outbreaks have occurred. Plague epidemics have not been known
in Western Europe since about 1721. In the nineteenth century there
were practically none west of Russia, and yet rats infested with fleas
are plentiful and ubiquitous.

[6] Hippocrates seems to have employed a method of auscultation.
Laennec, the father of modern auscultation, says: "*Ippocrate avait tenté
l'auscultation immédiate.*"

died on the eleventh day. The headache, the sweating, the delirium, and the eruption, the onset and length of the disease, are all as one would expect them to be in a severe case of typhus. The question of the diagnosis turns largely on the nature of the eruption, and this depends, of course, entirely upon the exact meaning attached to the words describing it. The significant expression is ἐξανθήματα μετὰ ἱδρῶτος ἐρυθρὰ σρογγύλα σμικρὰ οἷον Ἴονθοι. The οἷον Ἴονθοι has been translated by Farr as meaning "like vesicles," and by de Mercy as "*semblable aux varices.*" Professor Gulick, who has been good enough to take an interest in our classical dilettantisms, advises as follows: "I can find no other occurrence of the word Ἴονθος in Hippocrates, so that it is impossible to check up on *his* use of it. From Aristotle (*Hist. Animal.*, V, 31), it is clear that Ἴονθοι (originally the root of a hair) could occur either with or without pus. In *Problem.* xxxvi, 3, he asks why they occur mostly on the face; and in xxxiv, 4, he says that 'excrescences' — literally, 'hail,' or knots on the tongue — are like Ἴονθοι (exactly the expression in Hippocrates). Galen (xii, 824, ed. Kühn) says that boils, like Ἴονθοι, come from the skin moistures (he calls them juices), and that they are either hard and crude, or inflamed; in the latter case, fever subvenes; and he then gives several prescriptions for their treatment." It is therefore pure conjecture to regard this as a case of typhus fever. Indeed we think this improbable, when it is considered that no other similar ones are mentioned.

The tenth case in the series, the Clasomenian, whom Ozanam regards as definitely a case of typhus, appears —

on careful reading of the original — more like a severe typhoid fever.

There is not, therefore, anywhere in Hippocrates a clinical description which can be definitely recognized as applying to typhus fever. The search is equally unsuccessful if one investigates the writings of other classical authors who are supposed to have described the disease. Euryphon, a contemporary of Hippocrates, a physician of the Cnidian School, is often cited in support of the antiquity of typhus fever. Galen (xvii, 1, ed. Kühn) says: "Such fevers Euryphon names 'livid' ($\pi o\lambda\iota\dot{a}s$), and he writes as follows: 'The fever becomes livid and attacks the top of the head ($\beta\rho\epsilon\gamma\mu\dot{o}s$) in recurrent attacks; the head aches, a pain seizes the bowels, and the patient vomits bile; when this pain holds him, it is not possible to see what ails him; the belly becomes dry and all the skin livid, and the lips as if he had eaten black mulberries; the whites of the eyes become livid, and the patient looks as if he were being strangled; when he suffers this less, he suffers changes in his condition very often.' " This again is obviously not typhus as we know it to-day, but the description might well serve as a vivid portrayal of a severe attack of cholera.

3

The oldest recorded epidemic often regarded as an outbreak of typhus is the Athenian plague of the Peloponnesian Wars, which is described in the Second Book of the *History* of Thucydides.

In trying to make the diagnosis of epidemics from

ancient descriptions, when the differentiation of simultaneously occurring diseases was impossible, it is important to remember that in any great outbreak, while the large majority of cases may represent a single type of infection, there is usually a coincident increase of other forms of contagious diseases; for the circumstances which favor the spread of one infectious agent often create opportunities for the transmission of others. Very rarely is there a pure epidemic of a single malady. It is not unlikely that the description of Thucydides is confused by the fact that a number of diseases were epidemic in Athens at the time of the great plague. The conditions were ripe for it. Early in the summer of 430 B.C. large armies were camped in Attica. The country population swarmed into Athens, which became very much overcrowded. The disease seems to have started in Ethiopia (ἐξ Αἰθιοπίας τῆς ὑπὲρ Αἰγύπτου), thence traveled through Egypt and Libya, and at length reached the seaport of Piræus. It spread rapidly. Patients were seized suddenly, out of a clear sky. The first symptoms were severe headache and redness of the eyes. These were followed by inflammation of the tongue and pharynx, accompanied by sneezing, hoarseness, and cough. Soon after this, there was acute intestinal involvement, with vomiting, diarrhœa, and excessive thirst. Delirium was common. The patients that perished usually died between the seventh and ninth days. Many of those who survived the acute stage suffered from extreme weakness and a continued diarrhœa that yielded to no treatment. At the height of the fever, the body became covered with reddish spots (ὑπέρυθρον, πελιτνὸν, φλυκταίναις μικραῖς καὶ ἕλκεσιν ἐξηνθηκός), some of which ulcerated.[7] When one of the very severe cases recovered, convalescence was often accompanied by necrosis of the fingers, the toes, and the genitals. Some lost their eyesight. In many there was complete loss of memory. Those who recovered were immune, so that they could nurse the sick without further danger. None of those who, not thoroughly immunized, had it for the second time died of it. Thucydides himself had the disease. After subsiding for a while, when the winter began, the disease reappeared and seriously diminished the strength of the Athenian state.

The plague of Athens, whatever it may have been, had a profound effect upon historical events. It was one of the main reasons why the Athenian armies, on the advice of Pericles, did not attempt to expel the Lacedæmonians, who were ravaging Attica. Athenian life was completely demoralized, and a spirit of extreme lawlessness resulted. Men no longer took trouble about what was estimated honor. As Thucydides expresses it: "They saw how sudden was the change of fortune in the case both of those who were prosperous and suddenly died, and of those who before had nothing but, in a moment, were in possession of the property of others." There was no fear of the laws of God or man. Piety and impiety came to the same thing, and no one expected that he would live to be called to account. Finally, the Peloponnesians left Attica in a hurry, not for fear of the Athenians, who were locked up in their cities, but because they were afraid of the disease. At the same time, the pestilence followed the Athenian

[7] φλύκταινα, a "rising" pimple, therefore unlike the "spot" of typhus.

fleet, which was attacking the Peloponnesian coast, and prevented the carrying out of the objectives for which their expeditions had been organized. Thus it is likely that the struggle between the two contending powers was influenced in its duration and in the swinging back and forth of the fortunes of war as much by the epidemic as by any generalship or force of arms.

The plague of Thucydides can be identified with no single known epidemic disease of our day. Haeser believes it to be more like typhus fever than any of the conditions familiar to us, and Hecker takes the view that it was typhus in a form from which it has been altered in the centuries that followed. The eruption was certainly not like that of typhus at the present time, but corresponds more nearly to that of smallpox. When all is said, we must conclude that the nature of the Athenian epidemic cannot be determined with certainty. The rapidity of spread in a crowded town of 10,000 relatively small buildings, with a tremendous influx of population, is consistent with many forms of epidemic disease. The onset, the immediate respiratory symptoms, the nature of the eruption, and the sequelæ might reasonably be interpreted as smallpox.

In trying to make a diagnosis of the Athenian plague, we must take seriously the suggestion made by Hecker that epidemic diseases may have been modified considerably in the course of centuries of alternating widespread prevalence and quiescence. One of the greatest achievements in the war which the medical sciences have waged against epidemic diseases is the discovery that, during times of quiescence in interepidemic periods, the potential agents of disease may smoulder in human carriers, in domestic animals, — especially rodents, — and in insects. And modern bacteriology has made considerable progress in revealing changes that take place in the characteristics of bacteria and virus agents in the course of their adaptation to different environments. In the typhus-fever group, these circumstances have been most particularly studied, and we already have knowledge of a number of varieties of typhus and typhus-like fevers which have developed within historic times, probably because of the passage of the virus through different varieties of rodents and of insects and through man. These are matters which we have discussed more precisely in another place.

Thus, in endeavoring to classify the plague of Athens in the fifth century B.C., we have to choose between typhus, bubonic and pneumonic plague, and smallpox.

There is, in our opinion, practically no reason for assuming that the disease in question was a variety of typhus. Whatever may be the difference of opinion about the words φλύκταινα or ἕλκεα it seems fairly certain that the eruption, unlike that of typhus, was raised and, later, vesiculated; and the sudden onset, prominently marked by the inflammatory symptoms of the upper respiratory tract and severe coughs, is also inconsistent with epidemic typhus as we know it. The necroses of the extremities do suggest typhus, but this symptom is not usually prominent except in winter epidemics in armies, and the Athenian disease began early during a hot summer. This seasonal factor is also against typhus. Moreover, careful scrutiny of other ancient evidence does not give us reason to believe that typhus was known or reliably described until long after this period.

Bubonic plague probably existed. It is quite certain that it was prevalent in the Near East and on the northern coast of Africa at least three hundred years before Christ, and in other places we have shown that the bubonic form, or a closely related condition, caused severe epidemics in Biblical times. But there is nothing whatever in the description of the Athenian plague by Thucydides which would give an indication that the *Bacillus pestis* or a similar organism, either in the bubonic or in the pneumonic form, could have caused this epidemic.

We are led to consider smallpox or a variety of smallpox as the most likely classification. Whether smallpox was prevalent in the world at this time or not has been much disputed. Littré believed that there was no positive evidence of this in ancient literature. On the other hand, Haeser cites passages in Susruta which seem to refer to a disease prevalent in ancient India which closely simulated smallpox, and Paschen accepts the evidence which has been advanced to show that smallpox existed in China as early as 1700 B.C. In general, there seems to be considerable unanimity on the part of learned writers that smallpox was absent from Europe during the Greek and Roman classical periods.[8] In spite of this, however, the

[8] It is assumed by some writers that smallpox was spread over Europe with the wandering Gothic and Germanic tribes, but this is more or less guesswork. It is definite that it was a common condition all through North Africa by the time of the sixth century A.D., and about the same

description of Thucydides seems to us to point directly
to a disease of this general type. This surmise is strengthened by the occurrence of another epidemic, described
by Diodorus Siculus, which attacked the Carthaginian
army besieging Syracuse in 396 B.C., less than forty years
after the outbreaks in Attica. Diodorus describes it as

---

period there was an epidemic in France, described by the Bishop of
Avranches and by Gregory of Tours, which was quite surely smallpox.
Rhazes, who wrote during the early part of the tenth century, describes
the disease accurately, and during his time it was widely distributed
throughout the Near East, which it is believed to have reached through
Arabia from Abyssinia during the "Elephant War" in the fourth century A.D. Later, it was carried by the Saracens into Spain, whence it
quite naturally penetrated into Europe.

By the year 1000, it was present in practically all the European nations and was again and again reintroduced from the East by returning
Crusaders. Indeed, it is likely that the sad fate of the army of Frederick Barbarossa was brought about by smallpox and not by force of
arms. The Mongolian invasion brought a new mass inoculation, as a
consequence of which the first pesthouses had to be built to shelter the
immense numbers of the sick. Eventually, this disease was regarded
as one of the inevitable trials of all men.

After the discovery of America, smallpox followed close on the
heels of the discoverers. In the conquest of Mexico and in the rapid
subjection of the powerful native tribes, the European was unquestionably assisted by his powerful allies the pestilences, to which the aborigines were as susceptible as children. Among these, smallpox was the
most effective. A Negro from the ship of Narvaez carried smallpox
ashore, and over 3,000,000 Indians are said to have died. Negro
slaves, indeed, quite possibly played a considerable rôle in the rapid
distribution of the pox throughout the new continent. By the middle
of the sixteenth century, it is clear that the entire world had become
infected with the virus.

The smallpox epidemics of the subsequent two centuries, recurring
whenever susceptible fuel had accumulated, were of an extent and
severity of which it is hard for us to form any conception at the
present time; and it is safe to say that this condition would still prevail, attacking each new generation, were it not for the single and
simple procedure of Jennerian vaccination.

follows: "First, before sunrise, because of the cold breezes
from the water, they had chills; in the middle of the day,
burning heat. During the first stage of the disease there
was a catarrh (κατάρρους); followed by a swelling in
the throat (τράχηλος); shortly after this, fever set in;
pains in the back and a heavy feeling in the limbs; next,
a dysentery and blisters (φλύκταινα) upon the whole
surface of the body." After this, some became delirious.
Death occurred on the fifth or sixth day in most cases.
Diodorus attributes the disease to the multitude gathered
together in one place, the dryness of the summer, and the
"hollow and marshy" nature of the place. There was an
enormous death rate; the siege had to be raised, and the
army dispersed. From an historical point of view, this
epidemic was of the greatest importance, because it meant
that less than one hundred years before the outbreak of
the Punic Wars, in which much of the early fighting took
place in Sicily, Carthage was prevented by this epidemic
from completely controlling Sicily with a powerful army
of occupation and well-organized naval bases. Rome had
the greatest difficulty in conquering the Carthaginians,
and decisive Carthaginian superiority in the earlier campaigns might well have resulted in supplanting the military and administrative civilization of Rome with the
commercial, Semitic culture of Carthage — an event
which would have modified profoundly all subsequent
history.[9] The disease as described by Diodorus — again
like the epidemic in Athens — corresponds about as closely

[9] It might have resulted in developing a commercial civilization like
our own several thousand years earlier.

as can be expected of ancient descriptions to the severe, confluent type of smallpox in which death on the fifth or sixth day is not exceptional.

It is interesting to note that a similar epidemic attacked both the Roman and the Carthaginian army in 212 B.C., when they met in battle at Syracuse, but the description of this outbreak is not sufficiently clear to permit diagnostic identification.

CHAPTER VII

A continuation of the consideration of diseases of the ancients, with particular attention to epidemics and the fall of Rome. We are still engaged in our search for evidences of the occurrence of typhus in ancient times

I

THE effects of a succession of epidemics upon a state are not measurable in mortalities alone. Whenever pestilences have attained particularly terrifying proportions, their secondary consequences have been much more far-reaching and disorganizing than anything that could have resulted from the mere numerical reduction of the population. In modern times, these secondary effects have been — to some extent — mitigated by knowledge which has removed much of the terror that always accompanies the feeling of complete helplessness in the face of mysterious perils.

In this respect, modern bacteriology has brought about a state of affairs which may exert profound influence upon the future economic and political history of the world. Some epidemic diseases it has converted from uncontrolled savagery into states of relatively mild domestication. Others it can confine to limited territories or reservations. Others again, though still at large, can be prevented from developing a velocity which — once in full swing — is irresistible. But even in cases where no

effective means of defense have as yet been discovered —
as, for instance, in influenza, infantile paralysis, and encephalitis — the enemy can be faced in an orderly manner,
with determination and with some knowledge of his probable tactics; still, no doubt, with terror, but at least without the panic and disorganization which have been as
destructive to ancient and mediæval society as the actual
mortalities sustained.

In earlier ages, pestilences were mysterious visitations,
expressions of the wrath of higher powers which came out
of a dark nowhere, pitiless, dreadful, and inescapable. In
their terror and ignorance, men did the very things
which increased death rates and aggravated calamity. They
fled from towns and villages, but death mysteriously
traveled along with them. Panic bred social and moral
disorganization; farms were abandoned, and there was
shortage of food; famine led to displacement of populations, to revolution, to civil war, and, in some instances,
to fanatical religious movements which contributed to
profound spiritual and political transformations.

The disintegration of the Roman power was a gradual
process brought about by complex causes. Although, at
the death of Honorius, in 423 A.D., Britain alone had
broken away from formal Roman control, the cracks along
which the eventual cleavages were to come had already
been well started. The edict of Caracalla, long before this,
had raised the inhabitants of the provinces to the dignity
of Roman citizenship, but in actuality the knights of
Rome had no more in common with the burghers of
Nicomedia or Augusta Trevirorum than a banker Republi-

can of Boston or New York to-day has in common with a farmer Democrat of Oklahoma. Gigantic bureaucracies were eating up the government, budgets were almost modernly unbalanced, and the barbarians, — already settled in the Empire, — immigrants in the modern sense,[1] were expressing their aspirations for political power by marching on the capital whenever farming ceased to pay. The Visigoths, settled by Theodosius south of the Danube, started a farmers' strike in 396 under Alaric, and were stopped from occupying Rome only by the payment of a large farm loan, then spoken of as a ransom. The Vandals and Suebi, in 405, took possession of Spain, crossed into Africa, and established a sort of Middle West, which could enforce its desires by controlling the grain supply.

The problem has been dealt with from every conceivable angle, for there is no greater historic puzzle than that of the disappearance of the ancient civilization — a disappearance so complete that not a spark from its embers shone through the barbaric darkness of several hundred years.[2] Historians have analyzed the causes according to the prejudices of their own varieties of erudition. Mommsen, Gibbon, Ferrero, deduce the disintegration of the state, with variations of emphasis, from a com-

[1] In support of this, we submit the fact that the final struggle for supremacy in Italy itself was between Genseric, the Vandal, and Ricimer, the Suebian, a situation not unlike the political contest in New York between Mr. O'Brien and Mr. La Guardia.

[2] The desolate completeness of the disappearance of every vestige of the ancient civilization and organization is vividly described in the first chapter of Funck-Brentano's *Le Moyen Age.*

bination of political, religious (moral), and sociological causes. Ferrero lays fundamental stress upon the "interminable civil wars which resulted from the efforts of later Rome to reconcile the two essentially different principles of monarchy and republican organizations." Some have attempted to explain the breakdown on a basis of agricultural failure (Simkhovitch, *Hay and History*); a few associated with this the influence of a formidable increase of malaria, which accelerated the desertion of the farm lands (Ross). Pareto (*Traité de Sociologie Générale*, Vol. II, Chap. XIII — "L'Équilibre Social dans l'Histoire") seems to us to have given the most reasonable analysis, in which, in an extraordinarily brief treatment, he correlates the many complex factors that were coöperatively active. But even he has failed to include any consideration of the calamitous epidemics which — sweeping the Roman world again and again during its most turbulent political periods — must have exerted a material, if not a decisive influence upon the final outcome.

We are far from wishing to make the error against which Pareto warns, *"d'envisager comme simples des faits extrêmement compliqués"*; and we do not mean to add to other one-sided views an epidemic theory of the Roman decline. But we believe that a simple survey of the frequency, extent, and violence of the pestilences to which Roman Europe and Asia were subjected, from the year one to the final barbarian triumph, will convince the unprejudiced that these calamities must be interpolated in any appraisal of the causes that wore down the power of the greatest state the world has known. Indeed, we are inclined to believe, from a consideration of the circumstances prevailing at that time, that it would be impossible to maintain permanently a political and social organization of the type and magnitude of Rome in the face of complete lack of modern sanitary knowledge. A concentration of large populations in cities, free communication with all other parts of the world, — especially Africa and the East, — constant and extensive military activity involving the mobilization of armies in camps, and the movement of large forces back and forth from all corners of the world — these alone are conditions which inevitably determine the outbreak of epidemic disease.[3] And against such outbreaks there was absolutely no defense available at the time. Pestilences encountered no obstacles. They were free to sweep across the entire world, like flames through dry grass, finding fuel wherever men lived, following trade routes on land, and carried over the sea in ships. They slowed down only when they had burned

[3] This is still entirely applicable to modern times. Experience in the cantonments of 1917 and in the sanitation of active troops convincingly showed that war is to-day, as much as ever, 75 per cent an engineering and sanitary problem and a little less than 25 per cent a military one. Other things being approximately equal, that army will win which has the best engineering and sanitary services. The wise general will do what the engineers and the sanitary officers let him. The only reason why this is not entirely apparent in wars is because the military minds on both sides are too superb to notice that both armies are simultaneously immobilized by the same diseases.

Incidentally, medicine has another indirect influence on war which is not negligible. There seems little doubt that some of the reckless courage of the American troops in the late war was stimulated by the knowledge that in front of them were only the Germans, but behind them there were the assembled surgeons of America, with sleeves rolled up.

themselves out — and even then, when they had traveled as slowly as did the plagues of Cyprian and Justinian, they often doubled on their own paths, finding, in a new generation or in a community with fading immunity, materials on which they could flame up again for another period of terror. As soon as a state ceases to be mainly agricultural, sanitary knowledge becomes indispensable for its maintenance.

Justinian died in 565. Charlemagne was crowned in 800. Between 600 and 800, Italy was the battleground of barbarian immigrants who were fighting for the spoils. Rome, in the ancient sense, had ceased to exist. The final collapse of its defensive energy corresponds, in time, with the calamity of the great pestilence which bears Justinian's name. And while it would not be sensible to hold this plague alone responsible, it can hardly be questioned that it was one of the factors — perhaps the most potent single influence — which gave the *coupe de grâce* to the ancient empire.

Moreover, the history of the preceding six hundred years furnishes any number of examples to show that, again and again, the forward march of Roman power and world organization was interrupted by the only force against which political genius and military valor were utterly helpless — epidemic disease. There is no parallel in recent history by which the conditions then prevailing can be judged, unless it is the state of Russia between 1917 and 1923. There, too, the unfettered violence of typhus, cholera, dysentery, tuberculosis, malaria, and their brothers exerted a profound influence upon political events.

But of this we shall have more to say presently. It was only the highly developed system of sanitary defense on the Polish and the southern fronts that prevented, during those years, an invasion — first of disease, misery, and famine; then of political disruption — from spreading across Europe. This statement may, perhaps, be debatable. But it is, at least, a reasonable probability.

At any rate, during the first centuries after Christ, disease was unopposed by any barriers. And when it came, as though carried on storm clouds, all other things gave way, and men crouched in terror, abandoning all their quarrels, undertakings, and ambitions, until the tempest had blown over.

We have searched in vain for evidences of typhus during this period — but the significance of epidemics for the decline of Rome is of such interest that we may be forgiven another brief digression.

2

There is relatively little information in the literature of the first century A.D. in regard to epidemics. In the reign of Nero (after 54 B.C.), a plague occurred which is described by Tacitus as "extraordinarily destructive" — though his text gives no clues from which a diagnosis can be made. In the cities of Italy, there raged a disease which was so severe that corpses were in all the houses, and the streets were filled with funeral processions. "Slaves as well as citizens died" (we quote from Schnurrer), "and many who had mourned a beloved victim died themselves with such rapidity that they were carried to the

same pyre as those they had mourned." Whether this
particular malady was confined to Italy or not, we have
no means of telling. But during the same period there
were a number of other epidemic diseases in the provinces,
one of which is described as "anthrax," and was probably
similar or identical with the infection known by this name
to-day, since it attacked cattle and horses as well as men.
According to some writers, it was this disease which,
occurring among the Huns about 80 A.D., started 30,000
of them, with 40,000 horses and 100,000 cattle, on their
westward wanderings (Johannes von Müller).

Throughout the first century, there were earthquakes,
famines, volcanic eruptions, and vaguely reported epidemics. However, the first pestilence of which we have
reliable accounts is that which is spoken of as the "Plague
of Antoninus" (or of Galen). This disease started in the
army of Verus, which was campaigning in the East in
165 A.D. According to Ammianus Marcellinus, the original
infection came from a chest in a temple which the soldiers
had looted. As the army returned homeward, it scattered
the disease far and wide, and finally brought it to Rome.
By this time, the infection had radiated into all corners of
the world, and before long had extended "from Persia to
the shores of the Rhine," even spreading through the
Gallic and Germanic tribes. The mortality in many of
the cities was such that, as Marcus Aurelius says, "corpses
were carried in carts and wagons." Orosius states that so
many people died that cities and villages in Italy and in
the provinces were abandoned and fell into ruin. Distress
and disorganization were so severe that a campaign against

the Marcomanni was postponed. When, in 169, the war was finally resumed, Haeser records that many of the Germanic warriors — men and women — were found dead on the field without wounds, having died from the epidemic. Marcus Aurelius contracted the disease and, recognizing the contagiousness of his affliction, refused to see his son.[4] He died on the seventh day, his illness aggravated by his refusal to take nourishment. Since this was in 180 A.D., at which time Galen's description, *Methodus Medendi*, was written, it is plain that the pestilence in Europe lasted at least fourteen years. There is no definite information of the approximate number of deaths, but there is no doubt about the fact that the mortality was so great that it completely demoralized social, political, and military life and created such terror that there were none who dared nurse the sick. Our authority for this is Ammianus Marcellinus. The temporary arrest of the epidemic in 180 lasted only nine years. Dio Cassius tells us that it broke out again under Commodus in 189. "There arose the greatest plague of any I know of. Often there were 2000 deaths a day at Rome." It appears that the later phases were even more deadly than the earlier ones.

The nature of this disease is uncertain. It is, as usual, more than likely that no single infection was responsible, but that a number of different ones were raging at the

[4] About the only thing that centuries and changing civilization, religions, and customs have not been able to alter is the biological law of affection.

same time. The most fatal of these, the one which gave
the epidemic its chief characteristics, was a condition which,
if not smallpox, was closely related to it. Indeed, the
epidemic of Antoninus seems to have closely resembled the
plague of Athens. Galen tells us that a majority of the
cases began with inflammations of the pharynx, fever, and
diarrhœa. On the ninth day, there was — in most cases
— an eruption which was sometimes pustular and sometimes dry. We are again faced with the difficulty of
accurately interpreting the words referring to the nature
of the exanthemata, but there is less uncertainty in connection with this disease than there was in descriptions of
the plague of Athens, in regard to the raised, often vesicular and pustular nature of the eruption. Haeser, whose
opinion in this matter we share, after reading the evidence, inclines to the belief that the epidemic was one of
smallpox, or of a disease closely related to the modern
form of the disease. This fact is rendered particularly
likely by the speed and extensiveness with which the
malady spread across the entire known world.

There can be little room for doubt that a calamity of
this kind, lasting for over a decade, during a political
period rendered critical by internal strife and constant
war against encircling hostile barbarians, must have had a
profound effect upon the maintenance of the Roman
power. Military campaigns were stopped, cities depopulated, agriculture all but destroyed, and commerce para-
lyzed.

Apart from the military and camp disease which at

brief intervals afflicted the frontier armies,[5] the Roman
world remained relatively free of great pestilences from
the time of Commodus until the year 250, a period when
the empire was entering into its turbulent, ever-increasing struggle against the barbarian inroads. The threat
became especially serious after the victory of the Goths
over Decimus at Forum Trebronii. There started at this
time a pandemic which is described, among others, by
Saint Cyprian — and is therefore often spoken of as the
epidemic of Cyprian. This disease, like the Athenian
plague, was said to have originated in Ethiopia, reach-
ing Europe after passing across Egypt. It lasted no less
than fifteen or sixteen years, during which it spread over
the entire known world "from Egypt to Scotland." It
swept over the same regions repeatedly, after intervals of
several years. Its contagiousness was extreme and, according to Cedrenus, it was transmitted not only by direct
contact, but indirectly — through clothing. Gregory of
Nyssa [6] and Eusebius have left records of the suddenness

[5] An indication of the frequent occurrence of camp disease in the
Roman armies is found in Vegetius's *De Re Militari*, dedicated to
Valentinian about 375 A.D. "An army must not use bad or marshy
water; for the drinking of bad water is like poison and causes plagues
among those who drink it." And, at the end of the chapter: "If a
large group stays too long during the summer or autumn in one place,
the water becomes corrupt, and because of the corruption, drinking is
unhealthy, the air corrupt, and so malignant disease arises which cannot
be checked except by frequent change of camp."

[6] In Gregory of Nyssa, the same plague is referred to as occurring
during the life of Gregorius Thaumaturgus. There is also a description in *Patrologia Græca, Gregorius III*, in which the symptoms are
given as follows: "When once the disease attacked a man, it spread
rapidly over all his frame. A burning fever and thirst drove men to the

of its appearance, and of its terrifying violence. In a city
of Pontus, in 256, it appeared after the gathering of a
great crowd in a theatre, as a punishment for the temerity
of the spectators in challenging Jupiter, in whose honor
the performance was given. In Alexandria, the mortality
was enormous. The speed of extension was favored by the
active warfare going on in many of the provinces. The
Germanic tribes were invading Gaul and the Near East.
The Far Eastern provinces were being attacked by the
Goths, and the Parthians were conquering Mesopotamia.
Terror was extreme, and phantoms were seen to hover
over the houses of those who were about to fall sick.
Saint Cyprian made many conversions to Christianity by
exorcising these evil spirits. Throughout the early Christian period, every great calamity — famine, earthquake,
and plague — led to mass conversions, another indirect
influence by which epidemic diseases contributed to the
destruction of classical civilization. Christianity owes a
formidable debt to bubonic plague and to smallpox, no
less than to earthquake and volcanic eruptions.

The nature of the plague of Cyprian is even more
difficult to determine than is that of the Athenian pestilence. Haeser believes that bubonic plague played a
dominant rôle, and bases this chiefly upon the seasonal
factor — that is, upon the reports that in Egypt successive
outbreaks began in the autumn and lasted until the very
hot weather in July. In the absence of any definite informa-

---

springs and wells; but water was of no avail when once the disease had
attacked a person. The disease was very fatal. More died than survived,
and not sufficient people were left to bury the dead."

tion of glandular swellings or buboes, however, this view is pure surmise. Cyprian describes the disease as beginning with redness of the eyes, inflammation of the pharynx and throat, violent diarrhœa and vomiting.[7] He mentions gangrene of the feet, paralysis of the lower extremities, deafness, and blindness. No skin eruption is described. One must assume again a synchronous prevalence of many diseases, among which forms of meningitis and probably acute bacillary dysenteries were frequent, but no specific diagnosis is possible from the symptoms observed by writers of the period.

Whatever the conditions may have been, their violence was so extreme that one cannot doubt their serious effects upon political and social development. A conception of the extreme distress may be obtained from the following, which we quote literally from Haeser: "Men crowded into the larger cities; only the nearest fields were cultivated; the more distant ones became overgrown, and were used as hunting preserves; farm land had no value, because the population had so diminished that enough grain to feed them could be grown on the limited cultivated areas." Even in the centre of Italy, large territories became vacant; swamps developed, and rendered unhealthy the formerly wholesome coast lands of Etruria and

[7] Cyprian's description in *De Mortalitate* is as follows: "The bowels, relaxed into a constant flux, use up the strength of the body. A fire, conceived in the marrow, ferments into wounds in the jaw [*fauces*]. The intestines are shaken with continual vomiting. The eyes burn with blood. Sometimes the feet or other parts of the limbs are cut off because of the infection of disease, [*causing*] putrefaction [*morbida putredo*]."

Latium. Hieronymus writes that the human race had been "all but destroyed," and that the earth was returning to a state of desert and forests.[8]

During the plague of Cyprian, according to Baronius, the Christian custom of wearing black as a color of mourning originated. It had been used before by Hadrian, who, says Schnurrer, wore black for nine days after the death of Plotina.

Between the pestilence of Cyprian and the next great pandemic, spoken of as the plague of Justinian, there occurred a succession of calamities — earthquakes, famines, and the severe, but relatively localized, epidemic diseases such as one would expect in an empire in which there was a constant movement of large armed forces and uninterrupted communication with the East and with the north coast of Africa. At the same time, the migration of agricultural populations to the cities had already produced a great crowding of people into small areas, without any of the indispensable safeguards of modern medicine.

In the reign of Diocletian and Maximian, a plague is described, without any specific symptomatology, by Cedrenus. Eusebius places this outbreak a little later, and

[8] In studying the long cyclic swings of history, one learns that the judgment of political, social, and other changes in human destinies must be based on periods of not less than two or three centuries. With our own experience, we can appraise only a fraction of the curve in the cycle of which we are a part, and we cannot look forward clearly unless we are trained and capable of looking backward to the beginning of the curve, at least two or three hundred years in the past. Do Mr. Roosevelt and his brain trusters realize this?

also speaks of a new disease — possibly anthrax — which affected thousands of people, appeared in the form of acute ulcerations and swellings of different parts of the body, and blinded many in whom it occurred. Numbers of domestic animals died at the same time. Disease and famine continued into the year 313.

There follows a period about which we have relatively little record, though it probably had its usual measure of disease. It is the period during which *Völkerwanderung* was in one of its most active stages. This phenomenon was like the impact of human waves from east to west. The movement may have been started when the Huns, or Hiong-nus, were pushed out of China, and wandered to the Caspian Sea. Impelled to move, possibly by disease,[9] they began to migrate westward. Their first collision was with the Alani, whom they scattered or carried along with them in a thrust against the Goths. The latter had wandered from the north along the river beds toward the Black Sea. Crowded out by Huns and Alani, the Goths fled into Roman territory, where they temporarily settled along the Danube.

By 406, a general movement of barbarian tribes — Suebi, Alani, Burgundi, and Vandals — was taking place into Italy, Gaul, and across the Pyrenees to Spain. According to Idatius, it was a period of war, famine, and pestilence. In 444, there was a terrible epidemic in Britain, which seems to have been in part responsible for the historically momentous conquest of Britain by the Saxons. Bæda, in his *Historia Ecclesiastica Gentis An-*

[9] Suggested by Schnurrer.

glorum,[10] states that Voltiger, hard pressed and in distress, called upon the Saxon chieftains, Hengist and Horsa, for assistance: "A terrible plague fell upon them, which destroyed so many that the living could barely suffice to bury the dead. They consulted what was to be done, and where to seek aid against the frequent incursions of the northern races [apparently their fighting forces were greatly depleted by the plague], and agreed to call in the Saxon nation from across the Sea. The Saxons arrived in 449, and acted as mercenary guards for the Britons." It requires little exercise of the imagination, therefore, to conclude that the history of the British Isles in all its subsequent developments of race, customs, architecture, and so forth, was in large part determined by an epidemic disease.

Eusebius tells of an epidemic which occurred throughout the Roman provinces and near Vienna (then known as Oræ Favianæ) in 455 and 456. It began with inflamed eyes, swelling and redness of the skin over the entire body, and it ended — sometimes fatally — on the third or fourth day, with severe pulmonary symptoms. It is impossible to say what this disease might have been — possibly general streptococcus infection, or a form of scarlet fever, with secondary streptococcic pneumonia.[11]

In 467, Rome itself suffered from a disease about which we know, from Baronius, only that it killed a great many people. In the immediately succeeding years, much

[10] *Beda Venerabilis, Opera Omnia*, Giles Edition of 1843, Vol. II, Book I, Chap. XIV and XV.

[11] A hæmolytic streptococcus pneumonia epidemic among troops occurred in one of the American cantonments in 1917.

scattered — but localized — epidemic disease occurred in the Gallic provinces; and in 477, when the Saxon King, Odoacer, reached Anjou on his way to Italy, a severe plague broke out among citizens and invaders alike. Shortly after this, a famine and plague in North Africa decimated the Vandals, thus preparing them for defeat by the Mohammedans.

Of great diseases there is no record during the ensuing fifty years, but in 526 occurred the great earthquake of Antioch, which was responsible for the death of several hundred thousand people.

This brings us to the greatest of all the pandemics that helped to undermine the ancient civilization — namely, that of Justinian, details of which we know very largely from the writings of Procopius.

The sixth century was a period of calamity rarely equaled in history. Seibel in his *Die Grosse Pest zur Zeit Justinians,* has thoroughly compiled the available information, and is the authority from which most subsequent writers quote. According to him, a succession of earthquakes, volcanic eruptions, — Vesuvius in 513 was one, — and famines preceded and accompanied the series of pestilences which wrought terror and destruction throughout all of Europe, the Near East, and Asia for over sixty years. Of the natural convulsions, the most destructive was an earthquake, followed by conflagration, which destroyed Antioch in 526, killed between 200,000 and 300,000 inhabitants, and frightened away most of the remainder. There also were earthquakes in Constantinople and in other cities of the East, as well as in many places in

Europe proper. Among others, there was a severe one in Clermont, then called Civitas Averna. A succession of floods and famines added to the general misery. The impoverishment, the displacement of populations, the agricultural disorganization and famine which attended these calamities must have contributed materially to the origin and spread of the pestilence. Modern experience has demonstrated this a number of times, when tidal waves, earthquakes, and floods have wrought similar havoc.

The great plague of Justinian began in Egypt, near Pelusium. The suggested Ethiopian origin is vague; there was a sort of ancient and traditional suspicion that disease usually came out of Ethiopia. Procopius writes of this: —

At this time [540], there started a plague. It appeared not in one part of the world only, not in one race of men only, and not in any particular season; but it spread over the entire earth, and afflicted all without mercy of both sexes and of every age. It began in Egypt, at Pelusium; thence it spread to Alexandria and to the rest of Egypt; then went to Palestine, and from there over the whole world; in such a manner that, in each place, it had seasonal occurrence. And it spared no habitations of men, however remote they may have been. And if, at times, it seemed as though it had spared any region for a time, it would surely appear there later, not then attacking those who had been afflicted at an earlier time; and it lasted always until it had claimed its usual number of victims. It seemed always to be spread inland from the coastal regions, thence penetrating deeply into the interior.

In the second year, in the spring, it reached Byzantium and began in the following manner: To many there appeared phantoms in human form. Those who were so encountered, were struck by a blow from the phantom, and so contracted the sickness. Others locked themselves into their houses. But then the phantoms appeared to them in dreams, or they heard voices that told them that they had been selected for death.[12]

Since Procopius himself believed these things, his account reflects the terrified helplessness and panic which spread with this pestilence.

Four months the plague remained in Byzantium. At first, few died — then there were 5000, later 10,000 deaths a day. "Finally, when there was a scarcity of grave-diggers, the roofs were taken off the towers of the forts, the interiors filled with the corpses, and the roofs replaced." Corpses were placed on ships, and these abandoned to the sea. "And after the plague had ceased, there was so much depravity and general licentiousness, that it seemed as though the disease had left only the most wicked."

Procopius devotes a number of paragraphs to a description which is our only clue to diagnosis: —

They were taken with a sudden fever: some suddenly wakened from sleep; others while they were occupied with various matters during the daytime. The fever, from morning to night, was so slight that neither the patients nor the physician feared danger, and no one believed that he would die. But in many even on the first day, in others on the day following, in others again not until later, a bubo appeared both in the inguinal regions and under the armpits; in some behind the ears, and in any part of the body whatsoever.

To this point, the disease was the same in everyone, but in the later stages there were individual differences. Some went into a deep coma; others into violent delirium. If they neither fell asleep nor became delirious, the swelling gangrened and these

[12] *De Bello Persico,* Chap. XXII.

died from excess of pain. It was not contagious to touch, since no doctor or private individual fell ill from the sick or dead; for many who nursed or buried, remained alive in their service, contrary to all expectations. Some of the physicians unacquainted with this disease and in the belief that the buboes were the chief site of the sickness, examined the bodies of the dead, opened the buboes and found a great many pustular places.

Some died at once; others after many days; and the bodies of some broke out with black blisters the size of a lentil. These did not live after one day, but died at once; and many were quickly killed by a vomiting of blood which attacked them. Physicians could not tell which cases were light and which severe, and no remedies availed.

Agathius, speaking of the year 558, describes the same disease at Byzantium and again mentions buboes and sudden death which usually occurred on the fifth day. It attacked all ages, but killed more men than women.

It is interesting to note that this epidemic displayed one of the characteristics so often referred to in modern epidemiology — namely, when the outbreaks began, the number of sick and the mortality were relatively slight, but both rose with appalling violence as the epidemic gathered velocity.

There can be little doubt that the pestilence of Justinian was mainly one of bubonic plague, but the references to the general eruption of black blisters in many cases indicate that smallpox of a very severe type participated. Whatever it was, its extent and severity were such that commentators like Haeser believe it to have exerted an influence upon the decline of the Eastern empire which historians have too often overlooked. In the course of

sixty to seventy years, a considerable part of the known
world was devastated by the disease. Cities and villages
were abandoned, agriculture stopped, and famine, panic,
and the flight of large populations away from the infected
places threw the entire Roman world into confusion.

Gibbon, speaking of this plague, says: "No facts have
been preserved to sustain an account or even a conjecture
of the numbers that perished in this extraordinary mortality. I only find that, during three months, five and at
length ten thousand persons died each day at Constantinople; and many cities of the East were left vacant, and
that in several districts of Italy the harvest and the vintage withered on the ground. The triple scourges of war,
pestilence and famine afflicted the subjects of Justinian;
and his reign is disgraced by a visible decrease of the
human species which has never been regained in some of
the fairest countries of the globe."

Procopius was an eyewitness of most of the events which
he describes. He was associated closely with Belisarius in
his campaigns, and occupied a position of sufficient importance to have the "inside" of what was going on in
Constantinople at the court. One may, therefore, assume
that his accounts of the turbulence of the period — wars,
political corruption, and pestilence — are not unduly
exaggerated. And since we have recently had a greater,
more widespread, and more destructive war than most
others of history, and since political corruption to-day is
probably quite as well developed and general as at any
time, it is a reasonable conjecture that it may have been
only our relative ability to control pestilence which has

preserved the modern world, for a time, from breaking up as did the empire of Justinian.

In studying, through the eyes of Procopius, the reign of Justinian, one obtains an extraordinarily vivid picture of the manner in which the three major agencies cooperated in bringing the empire to its knees. Justinian was making a final effort to restore the imperial world power. Wars with Persia, wars against the Vandals in Africa and against the Goths in Italy, armies to maintain on all fronts, in widely separated parts of the world, strained the resources of the government to their utmost. Everywhere the ring of defense was being pushed back by ever-increasing hordes of barbarians, who had by this time learned much of the art of war and of organization from their former overlords. Internal insurrections, as at Byzantium in 532, threatened the rear. Treachery and graft weakened the administrative power at court. And superimposed upon these almost, perhaps entirely insuperable difficulties was the pestilence, — sweeping from east to west, north to south, again and again, for almost sixty years, — killing, terrifying, and disorganizing.

The plague lasted until 590, or a little later. Between 568 and 570, most of Italy was conquered by the Lombards, who, as Cunimund, another barbarian, said, "resemble in figure and in smell the mares of the Sarmatian plains." The power and the grace and the administrative logic that once were Rome had died.

CHAPTER VIII

*On the influence of epidemic diseases on political and
military history and on the relative unimportance of generals. This, we promise, is the last serious digression from
our main theme*

IF it were not for the fact that so many utterly uninterested people die of disease or are killed in them, wars
would not be taken so seriously. It is of course true that
rapacity for territory, commercial rivalry, and all other
expressions of that avarice which is as instinctive to the
human species as the sexual and intestinal functions, have
always been present as the underlying causes of war. But
it is doubtful whether these more or less realistic reasons
would fulminate to the actual point of explosion as often
as they do if mankind did not, in spite of repeated demonstration, obstinately harbor a totally erroneous conception
of what actually constitutes a war in terms of experience.
It is not, of course, the propaganda of glory, the *dulce est
pro patria mori*, and so forth, that influence men so
deeply. These and similar "residues" (we hope we
are correct in our Pareto [1]) are only moderately effective
rationalizations of more fundamental impulses. Much
more deeply significant are the boredom with the un-

[1] See *An Introduction to Pareto*, by Homans and Curtis, or write
a letter to Professor Lawrence J. Henderson.

utterably dull peace-time occupations of most people, and the childish but universal delight men take in playing soldiers. Until they actually suffered from dirt, lousiness, fatigue, terror, disease, or wounds, most men enjoyed the last war. Think of the man who has lived meagrely in a frame house on the outskirts of Somerville or Weehawken, and for ten years — except for two weeks in August — has regularly caught the eight-fifteen, spent the rest of the day floorwalking, and then caught the six-twenty back to what he came from in the morning! Think of his feelings of release and self-satisfaction when he is marching up Broadway behind the band, between files of cheering garment workers. Think of his pride in a renewed manhood, standing guard at dawn or lying behind a pile of sandbags pot-shooting his fellow man, or drinking beer with his comrades — knowing that the world approves him as a hero, and that his family has the government to look out for it forever and ever!

But beyond the release from boredom there is the joy in uniforms which stimulates war. The instinct for fancy dress is hard to kill, as anybody knows who has been in a town where the Mystic Knights or the Shriners or the Red Indians were holding a convention; or even in Boston, when the Ancient and Honorables are blocking traffic on Beacon Hill. And, further, there is the applause of the women, — not women in general, but each man's own women, — who, as instinctively as the men like to play soldiers, have the hereditary longing to glorify the brave brutalities that their heroes write home about: "I threw a hand grenade into a dugout, and blew up

six Germans. I 'm going to be kissed by the general."
"Is n't he wonderful? Just a big, brave boy!" One can
hear the devil's grandmother, adoringly watching him
turn a squealing sinner on the spit, saying: "Oh, Beelzebub
— you 're nothing but a great big boy!"

We might expostulate on the minor causes of war in
a more convincingly thorough manner if we were writing
a tract for a peace foundation instead of the biography
of a disease. But since we are primarily interested in the
subject of typhus fever, we cannot give too much space to
these matters. The point is that war is visualized — even
by the military expert — as a sort of serious way of playing soldiers. In point of fact, the tricks of marching and
of shooting and the game called strategy constitute only
a part — the minor, although picturesquely appealing
part — of the tragedy of war. They are only the terminal
operations engaged in by those remnants of the armies
which have survived the camp epidemics. These have often
determined victory or defeat before the generals know
where they are going to place the headquarters' mess.

To the average professional officer, the military doctor
is an unwillingly tolerated noncombatant who takes sick
call, gives cathartic pills, makes transportation trouble,
complicates tactical plans, and causes the water to smell
bad. Of course, he is useful after an action, to remove the
débris, but otherwise he is almost, if not quite, a positive
nuisance. There was a tempest of respiratory diseases and
the threat of enteric fever in the Second American Army
at the end of the war. The inspector general, Colonel O.,
neither knew nor cared about that. He reprimanded a

weary chief sanitary inspector for saluting him with one hand in his pocket. We pitied this poor gentleman when we thought of all the buttons that were off and the puttees wrongly adjusted among a hundred thousand men. How he suffered and toiled! The same sanitary officer was trying to locate water points for the advancing troops in September 1918. "You don't exist for me," said Colonel H. of the Engineers. "You are not in the Tables of Organization." Occasionally there is a great soldier who knows, like General Bullard. He stands out by contrast. However, this may seem like spleen. But not at all; it leads up to our theme that soldiers have rarely won wars. They more often mop up after the barrage of epidemics. And typhus, with its brothers and sisters, — plague, cholera, typhoid, dysentery, — has decided more campaigns than Cæsar, Hannibal, Napoleon, and all the inspector generals of history. The epidemics get the blame for defeat, the generals the credit for victory. It ought to be the other way round — perhaps some day the organization of armies will be changed, and the line officer will do what the surgeon-general lets him do. Among other things, this plan would remove about 90 per cent of the expenses of the pension system.

Before we go on to the special military exploits of typhus, it may be interesting to discuss the decisive influence of disease upon battle in a more general manner, and so justify our contentions with a few facts.

The difficulty is not to find evidence, but to select from the dreadful abundance. Von Linstow, a military surgeon of the Prussian army, who thought along similar lines, has culled the literature for some of the most enlightening examples in common historical records. We cite freely from his studies, and from the writings of historians and military surgeons who have accompanied great armies in campaigns.

Herodotus, in the Eighth Book of his *History*, tells us about the saving of Greece by λοιμός (possibly plague and dysentery), when Xerxes entered Thessalia with an army estimated at about 800,000 men. Soon after Greek territory was entered, supplies began to fail, and disease stepped upon the heels of undernourishment and hardship. The campaign was abandoned, and the Persian king swept back into Asia with less than half a million followers.

It was the plague of Athens which laid low for a time the power of Athens on land. In the second year of the disease, 300 knights, 45,000 citizens, and 10,000 free men and slaves died. Pericles himself succumbed, and the Lacedæmonians were left free to roam over the peninsula.

That the sieges of Syracuse by the Carthaginians in 414 and 396 B.C. were relieved by a disease probably identical with that of Athens is likely. There is no telling what might have been the outcome of the Punic Wars and of the future power of Rome if Hannibal had found his fleet and armies firmly established in Sicily.

In the civil struggles of Rome, in 88 B.C., the victory of Marius was decided by an epidemic which killed 17,000 men in the army of Octavius.

In 425 A.D., the Huns gave up their otherwise unimpeded advance upon Constantinople because a plague of unknown nature decimated their hordes.

What might have been the future of the power of the Saracen Empire if the King of Abyssinia had not been turned back from Mecca by the "sacred fire," no one can tell. This was what is commonly spoken of as the "Elephant War." The Abyssinian army of 60,000 men was completely disorganized by the ravages of a disease which, in description, sounds either like a severe form of small-pox or like a combination of erysipelas and general staphylococcus infection.

That the Crusades were turned back by epidemics much more effectively than they were by the armed power of the Saracens can hardly be questioned. The history of the Crusades reads like the chronicle of a series of diseases, with scurvy as potent as infections. In 1098, a Christian army of 300,000 men besieged Antioch. Disease and famine killed so many and in such a short time that the dead could not be buried. The cavalry were rendered useless within a few months by the death of 5000 of their 7000 horses. Nevertheless, the city was captured, after a nine months' siege. On the march to Jerusalem, the hosts were accompanied by an enemy more potent than the heathen. When Jerusalem was taken, in 1099, only 60,000 of the original 300,000 were left, and these, by 1101, had melted to 20,000.

The story of the second Crusade, led by Louis VII of France, is sadly similar. Of half a million men, only a handful — most of them without horses — managed to get back to Antioch, and few returned to Europe.

Antioch seems to have been the spot where all the Christian armies were ambushed by pestilence. Error in the road taken beyond this city, through the treachery of a Turkish guide, led the crusading army of 1190 into the desert. Famine, plague, and desertions reduced an army of 100,000 to a mere 5000.

The fourth Crusade, under the Doge of Venice and Baldwin of Flanders, never reached Jerusalem because of a dreadful outbreak of bubonic plague which started during the hottest part of the summer, soon after the Crusaders left Constantinople.

When Frederick II of Germany took ship at Brindisi in 1227, dysentery came aboard with his army; the fleet turned back when the Emperor himself was taken sick, and the expedition was a flat failure.

Scurvy is not an infectious disease and has no proper place, therefore, among the relatives of typhus fever, whose influence on history we are discussing. However, it was an almost constant menace to armies whenever the food supply ran low or became restricted. Under such circumstances, which were common in besieged cities and during long marches through devastated territories, scurvy not infrequently became decisive in itself or so weakened large bodies of men that subsequent infectious disease found them without normal powers to resist. In this way it was often a powerful ally of our disease. We have no intention of further digressing from our main theme into the interesting military history of scurvy, but cite a single episode only, to illustrate the formidable influence of scurvy in determining the outcome of campaigns.

Until the first Friday in Lent of 1250, the crusading army of Saint Louis was reasonably holding its own against the Saracens. Shortly after this, Joinville tells us, "the host began to suffer very grievously." He attributes the nature of the illness to the stench of dead bodies and to the eels from the river that "ate the dead people, for they are a gluttonous fish." The disease was, without question, scurvy: "There came upon us the sickness of the host, which sickness was such that the flesh of our legs dried up, and the skin upon our legs became spotted; black and earth colour like an old boot; and with us who had this sickness, the flesh of our gums putrefied; nor could anyone escape from this sickness but he had to die. The sign of death was this, that when there was bleeding of the nose, then death was sure." The Turks at about this time managed to blockade the river against the supply ships, fresh food became still more scarce, and many of the leaders fell sick. "The sickness began to increase in the host in such sort, and the dead flesh to grow upon the gums of our people, that the barber surgeons had to remove the dead flesh in order that the people might masticate their food and swallow it. Great pity it was to hear the cry throughout the camp of the people whose dead flesh was being cut away; for they cried like women labouring of child." The disease made prompt retreat imperative, and the King decided upon a desperate effort to break through the Saracen blockade. Failure, defeat, and the capture of the King with all his knights followed.

On the second attempt, Louis got no farther than

Tunis, where he and his son, the Duc de Nevers, died of dysentery on August 3 and August 25, 1270.

A curious disease that cannot be precisely classified destroyed the army of Frederick Barbarossa in Rome in 1157. It is described by Kerner and also by Lersch. It might have been typhus, for it began with severe headaches, pain in the limbs and abdomen, heat, chills, and delirium. Many died within a few days. The mortality was so high and the terror so great that on August 6 of 1167, four days after the plague began, the army burned their tents and started northward. Rome was abandoned, and the greater part of the host perished on its march.

The centuries of struggles between Spain and France were again and again decided by disease. Philip III of France was turned back from his campaign into Aragon in 1285 by a plague of uncertain nature that killed large numbers of the soldiers, most of the officers, and, eventually, the King himself. In the subsequent military history of Spain, typhus itself played a devastating rôle, to which we shall have occasion to return in a later chapter.

In 1439, on October 1, the German Emperor, Albrecht, reached the walls of Bagdad. By the thirteenth of the same month, the Emperor was dead and the army in retreat, defeated by dysentery.

The rôle played by the sweating sickness in England during the reign of Henry II we have elsewhere described. We have also discussed the influences which the epidemic of syphilis had upon the campaign of Charles VIII of France against Naples; and in another place we shall speak of the typhus epidemic which, in 1528, decided whether France or Spain was to dominate the European Continent.

In the sixteenth century, the story is the same in principle, and though typhus and plague now begin to be cast for the leading rôles, dysentery, typhoid, and smallpox no doubt contributed their share. The siege of Metz by Charles V was raised by scurvy, dysentery, and typhus, and the army retreated from the city after 30,000 men had died.

One of the earliest really decisive typhus epidemics was that which dispersed the army of Maximilian II of Germany, who was preparing with 80,000 men to face the Sultan Soliman in Hungary. In the camp at Komorn, in 1566, a disease broke out which was undoubtedly typhus. It was so violent and deadly that the campaign against the Turks was given up. The significance of this episode for the permanent establishment of the disease in Southeastern Europe is discussed in another chapter.

The Thirty Years' War was in all its phases dominated by deadly epidemics. To follow them in detail would be to write the history of this war over again, for the pestilences roamed the Continent in the trains of the armed forces. There is one episode, however, which deserves particular mention, because typhus, single-handed, defeated both armies before they could join battle. In 1632, Gustavus Adolphus and Wallenstein faced each other before Nuremberg, which was the goal of both armies. Typhus and scurvy killed 18,000 soldiers, whereupon both the opposing forces marched away in the hope of escaping the further ravages of the pestilence.

It is not impossible that the fate of Charles I was sealed by typhus fever. In 1643, Charles was opposed at Oxford by the Parliamentary army under Essex, each general commanding about 20,000 men. The King was forced to give up his plan of advancing upon London by an epidemic of typhus fever which ravaged both armies.

In 1708, the Swedes, having their own way in Southern Russia, completely lost the fruits of their hard-fought battles and were rendered helpless by an outbreak of plague.

In November of 1741, Prague was surrendered to the French army because 30,000 of the opposing Austrians died of typhus.

Frederick the Great, victorious over the troops of Maria Theresa, was forced out of Bohemia when violent dysentery attacked his troops.

The outcome of the French Revolution was to some extent decided by dysentery. In 1792, Frederick William II of Prussia, with Austrian allies, a total strength of 42,000 men, was marching against the armies of the Revolution. Dysentery, the Red, decided in favor of *Liberté*, *Égalité*, and *Fraternité*, and with only 30,000 effectives remaining, the Prussians retreated across the Rhine.

The establishment of the Haitian Republic, though usually attributed to the genius of Toussaint l'Ouverture, was actually brought about by yellow fever. In 1801, Napoleon sent General Leclerc with 25,000 men to Haiti to put down the revolt of the Negroes. The French troops landed at Cap Français, defeated Toussaint, and drove him into the interior. The Negro army was rallied and reorganized by Dessalines, but could not have successfully opposed the well-disciplined and well-equipped French troops had not an epidemic of yellow fever disorganized the invader. Of 25,000 Frenchmen, 22,000 died. There were only 3000 left to evacuate the island in 1803.

Even the greatest general of them all, Napoleon, was helpless when pitted against the tactics of epidemic disease. We have accounts of the Russian campaign from Larrey. But records of more specific value for our subject are those of the Chevalier J. R. L. de Kerckhove (*dit* de Kirckhoff*), a corps surgeon of the army of invasion, who — on the title-page of his book — signs himself *"Membre de la plupart des Académies savantes de l'Europe."* The army of upward of half a million men was mobilized in cantonments which extended from Northern Germany to Italy. Until the main bodies were assembled, there was little sickness, and the hospitals established at Magdeburg, Erfurt, Posen, and Berlin had few patients. Kerckhove describes the miserable conditions encountered after the entry into Poland. He was shocked by the poverty, wretchedness, and slavishness of the people, and the contrast of the conditions here found with those prevailing in other European countries. The villages consisted of insect-infected hovels; the army was forced to bivouac. Nutrition was bad; the days hot and the nights cold. New hospitals were now established at Danzig, Königsberg, and Thorn, because of the rapidly increasing sick rates, at this time largely due to respiratory infections, including pneumonia and throat anginas — probably diphtheria. Typhus cases began to appear in small numbers at about the time that the Niemen was crossed, on June 24. In Lithuania, huge forests and wretched roads were encountered; towns and villages had been burned by the Russians; there was little shelter, and less food. The water was bad, the heat intense, and the disease rate — now largely dysentery, enteric fevers, and typhus — became formidable. After the battle of Ostrowo, in late July, there were over 80,000 sick. The army corps to which Kerckhove was attached was reduced to less than half of its original 42,000 men by the time the River Moskva was reached in early September. An enormous number of wounded — over 30,000 — resulting from the battle, fought near the river, further rendered the task of the medical officers an almost impossible one. By September 12, typhus and dysentery were becoming more and more intense. Moscow was entered on September 14. It was at this time a city of 300,000 people, but most of the population had fled before the French army entered. On the fifteenth, fires were started, first at the Bourse, then all over the city — set, presumably, under orders of Governor Rostoptchin, by liberated criminals who had been furnished with sulphur torches. Moscow contained a number of well-equipped hospitals, but these were soon filled with the sick and wounded, and since so large a part of the city was either in ashes or destroyed by bombardments, the thoroughly infected troops were crowded in unsatisfactory shelters and camped outside the city. Stores of food had been almost completely destroyed by the Russians.

From now on typhus and dysentery were Napoleon's chief opponents. When the retreat from Moscow was begun, on October 19, there were not more than 80,000 men fit for duty. The homeward march became a rout, and the exhausted and sick troops were constantly harassed by the pursuing enemy. The weather grew intensely cold, and a large number — exhausted by sickness and fatigue — were frozen. In early November, when Smolensk was reoccupied, only 2000 of the cavalry were left, and there were about 20,000 patients in the hospitals of the city. Many typhus patients were left behind in Smolensk, which was evacuated on November 13. The disastrous crossing of the Beresina, in which Larrey was saved only by the grateful affection of soldiers who passed him over their heads across the bridge, cost the army an enormous number — not precisely recorded, but estimated at 40,000 men. While typhus remained the predominant disease, dysentery and pneumonia were also increasing. Fifteen thousand men are said to have been frozen on the way to Vilna, and when this city was reached, on December 8, the magnificent army had shrunken to 20,000 sick and disheartened men. Of the Third Army Corps commanded by Marshal Ney, only twenty men remained. In Vilna the hospitals were crowded, the men lay on rotten straw in their own refuse, hungry and cold, without care. They were driven to eat leather and even human flesh. The diseases, especially typhus, spread through all the cities and villages of the surrounding country. At one time, in December, the sick that had been evacuated to Vilna had accumulated to the number of 25,000. By the end of June, 1813, only 3000 of these remained alive. The vestiges of the army which escaped from Russia were almost without exception infected with typhus.

It is suggested by de Kerckhove, whose book testifies to a lively interest in the strategy of his great chief, that if Napoleon had been content to occupy Poland and attend to reorganization, including sanitary control, his campaign might have been a success and his power permanently established.

It is perhaps the greatest testimony to the genius of Napoleon that, after this disastrous failure, he was again — in 1813 — able to raise a new army of 500,000 men. These were mostly, for lack of available adult man power, young recruits, particularly suitable fuel for epidemic disease. By the time his new army faced the allies at Leipzig, preliminary battles at Bautzen, Dresden, and Karlsbad, together with disease, had reduced his forces to little more than 170,000 men with which to face 200,000 allies.[2] It is hardly debatable that the power of Napoleon in Europe was broken by disease more effectively than by military opposition or even by Trafalgar.

As far as the Crimean War is concerned, it is not possible to deduce the results of the struggle from disease, since the opposing armies suffered almost equally and disastrously from cholera, typhus, dysentery, and the lesser epidemic afflictions of armies. Nevertheless, this

[2] Von Linstow states that 105,000 were lost for service by battle casualty; 219.000 by disease.

war is of unusual interest for our theme, because there are available unusually accurate records which demonstrate how much more destructive than the clash of armed conflict is the power of disease. We have reliable accounts of the army epidemics of this war from Jacquot's *Du Typhus de l'Armée d'Orient*, and Armand's *Histoire Medico-Chirugical de la Guerre de la Crimée*. There were two separate typhus outbreaks — one which started in December 1854, the other in December of the following year. The disease began among the Russians, then attacked the British and the French, penetrated into Constantinople, there spread to the fleets and the merchant ships, and was distributed in all directions throughout Russia and Turkey. In 1855, after the battle of Alma, a severe cholera epidemic began which lasted through to April 1856. At the time of the greatest violence of the various diseases, 48,000 men were removed from the ranks by sickness within four months — or at the rate of 12,000 a month. According to Armand, the French sent something over 309,000 men east. Of these, 200,000 were hospitalized — 50,000 by wounds and 150,000 by disease. The following table, which we take from Von Linstow, summarizes the conditions which prevailed from 1854 to 1856.

	Wounded	Died of Wounds[3]	Sick	Died of Disease
French	39,869	20,356	196,430	49,815
English	18,283	4,947	144,390	17,225
Russians	92,381	37,958	322,097	37,454

[3] Including men lost in battle, and so forth.

CHAPTER IX

On the louse: we are now ready to consider the environment which has helped to form the character of our subject

I

THE formula for writing biographies of individual men and women has, as we have seen, been thoroughly worked out. Apart from the recent introduction of psychoanalytical methods and a little libido, it has remained more or less the same since Plutarch. In writing the biography of a protoplasmic continuity like typhus, it becomes necessary to develop a new formula. While, on the one hand, we can avoid many of the keyhole indiscretions of the Strachey, Ludwig, Maurois school, we are — in this instance — forced to give considerable space and attention to other unpleasant subjects more likely to repel than to attract. For typhus spends prolonged and, for its survival, essential phases of its existence within the bodies of lice, fleas, and rats. There may be other hosts not yet determined. But of these we know; and we must, therefore, follow our virus through these phases and endeavor to get the point of view of the fellow creatures that, though regarded with loathing by the superficial, are sufferers even as we are, and quite as innocent of intentional malice. For though we acquire the disease from them, they get it from each other and from us. So there would seem to be as much to be said on one side as on the other.

Obviously it is much more difficult to present the louse's point of view in its relationship to man than to elucidate the influence exerted, let us say, upon Chopin by George Sand, or upon Mark Twain by the respectable relations of Elmira, New York. We cannot, therefore, dismiss the matter with a brief scientific description of the sojourn of typhus in the louse. To achieve our purpose, though it may again delay the actual consideration of typhus itself for another chapter, we must endeavor to present the case of the louse in the humane spirit which a long intimacy has engendered in us. For one cannot carry pill boxes full of these little creatures under one's sock for weeks at a time without developing what we may call, without exaggeration, an affectionate sympathy; especially if one has taken advantage of them for scientific purposes and finds each morning a corpse or two, with others obviously suffering — crawling languidly, without appetite, and hardly able to right themselves when placed on their backs.

We advise the reader who is impatient to press through to typhus fever to skip this chapter, therefore, since it will occupy itself mainly with the louse. But to those who are inclined to criticize us for being excessively discursive, we may state that we are following the distinguished model of Pierre Beyle, whose footnotes are four times as extensive as his text.

The louse is foremost among the many important and dignified things that are made the subjects of raucous humor by the ribald. Despite the immense influence of this not unattractive insect upon the history of mankind, it is given, in the *Encyclopædia Britannica*, two thirds of a column — half as much as is devoted to "Louth, a maritime county in the province of Leinster," one fifth as much as is allowed for Louisville, Kentucky. This creature, which has carried the pestilence that has devastated cities, driven populations into exile, turned conquering armies into panic-stricken rabbles, is briefly dismissed as "a wingless insect, parasitic upon birds and mammals, and belonging, strictly speaking, to the order of Anoplura."

The louse shares with us the misfortune of being prey to the typhus virus. If lice can dread, the nightmare of their lives is the fear of some day inhabiting an infected rat or human being. For the host may survive; but the ill-starred louse that sticks his haustellum through an infected skin, and imbibes the loathsome virus with his nourishment, is doomed beyond succor. In eight days he sickens, in ten days he is *in extremis*, on the eleventh or twelfth his tiny body turns red with blood extravasated from his bowel, and he gives up his little ghost. Man is too prone to look upon all nature through egocentric eyes. To the louse, *we* are the dreaded emissaries of death. He leads a relatively harmless life — the result of centuries of adaptations; then, out of the blue, an epidemic occurs; his host sickens, and the only world he has ever known becomes pestilential and deadly; and if, as a result of circumstances not under his control, his stricken body is transferred to another host whom he, in turn, infects, he does so without guile, from the uncontrollable need for nourishment, with death already in his own entrails. If only for his fellowship with us in suffering, he should command a degree of sympathetic consideration.

The louse was not always the dependent, parasitic creature that cannot live away from its host. There were once free and liberty-loving lice, who could look other insects in their multifaceted eyes and bid them smile when they called them "louse." But this was even longer ago than the Declaration of Independence, for it took the louse many centuries to yield up its individualism.

It was so long ago that we have no records of any neolithic or even Neanderthal louse from which we can trace a clear line of descent. Indeed the ancestral problem has remained an extraordinarily difficult one. Many erudite students — preëminently Enderlein — have been inclined to derive the Siphunculata or sucking lice from the Rynchota or true bugs, largely on the basis of similarities of the mouth parts. But this idea is rejected as truly preposterous by Professor Handlirsch [1] and his followers, who

[1] Handlirsch (*Die Fossilen Insekten*) says: —

"*Was die zuerst genannte Gruppe (Siphunculata) anbelangt, so wurde sie ihrer, saugenden Mundteile wegen von vielen Autoren den Hemipteroiden angegliedert, wobei aber nicht berücksichtigt wurde, dass diese saugenden Mundteile absolut nicht von jenen der Schnabelkerfe abstammen können, weil sie in manchen Punkten noch ursprünglicher sind, so zum Beispiel in den nicht zu einer Rüsselscheide verwachsenen, manchmal noch frei erhaltenen Tastern des dritten Kieferpaares usw. Alle diese Tatsachen wurden von mir in einer gegen Enderlein gerichteten Schrift (Zool. Anz. 1905, 664) wohl hinlänglich erörtert, und ich kann mich hier damit begnügen, noch einmal hervorzuheben, dass sich die Siphunculatenmundteile nur von einem kauenden Typus ableiten lassen und sich ganz eng an jene der Mallophagen anschliessen. Nachdem nun auch in Bezug auf die übrige*

trace the descent of our lice from the fur- and feather-
eating Mallophaga (bird lice) for reasons unquestionably
well founded upon considerations far too intricately technical for superficial discussion. We could not do justice to
a subject so fundamental without extensive citation from
the works of specialists. We desire merely to indicate that
this problem of ancestry has led to dissension among louse
scholars, on occasion not entirely without passion, though,
unlike the question of the descent of man, it has not involved religious feelings.

The opinion of the learned Professor Handlirsch appears to be the one most generally favored among louse
scholars. Modern lice consist of two closely related varieties: the biting lice, or Mallophaga; and the sucking
lice, or Anoplura. These orders are parasitic developments,
probably, of the ancient group of pre-cockroaches, from
which are also derived our present cockroaches and termites. The pre-cockroaches, or Protoblattoidea, are fossil
forms of the upper Carboniferous period, and too far

*Morphologie eine weitgehende Übereinstimmung zwischen blutsau-
genden und pelzfressenden Läusen besteht, liegt es allzu nahe, erstere
von letzteren abzuleiten, beziehungsweise durch Vermittelung der Cor-
rodentien (Psociden) von der Blattoidenreihe. Diese Anschauung ent-
hebt uns der gewiss misslichen Nötigung, zu einem so unnatürlichen
und unlogischen Auskunftsmittel zu greifen, wie es eine Ableitung der
Pediculiden von der Wurzel des Hemipteroidenstammes wäre, denn
wir müssten in diesem Falle bis in das Palaeozoikum hinabsteigen, wo
es bekanntlich noch keine Säugetiere gab, auf denen ausschliesslich die
Pediculiden leben können. Für die Ableitung der Pediculiden von
Mallophagen ist übrigens in neuerer Zeit, gleichzeitig aber ganz un-
abhängig auch N. Cholodkowsky auf Grund der Embryonalentwicke-
lung dingetreten. Hoffentlich gelingt es unseren vereinten Bemühungen
doch endlich auch Enderlein von seiner Ansicht über die engen Be-
ziehungen zwischen Pediculiden und Hemipteroiden abzubringen."*

removed to concern us. Our own companions, the blood-sucking ones, are probably derived from the fur-scavenging insects, through the Psocidæ or Corrodentia — small winged or wingless creatures, the best known representatives of which are our common book lice. The latter group are not the direct ancestors of the louse, but spring with them from a common stem. The conditions are analogous to the relationship of the higher apes and man — a kinship that is too often misunderstood as a direct descent or ascent (however one looks at it), like the rungs of a ladder, rather than, more properly, like twigs of the same bush.

Ancestral origin from the same stock may be both upward and downward. In the case of the louse, we know relatively little about the matter, since we must judge from anatomical data alone; and the evolution of purely parasitic from free-living forms would seem to be a downward rather than an upward development. In the case of man, the relationship with the monkeys is surely much closer than that of the lice with the Psocidæ. The anatomical and blood-chemical similarities are exceedingly close ones and — being the arbiters of appraisal — we assume that we are the higher forms, since we include mental and spiritual qualifications, without really knowing much of these attributes among the apes. A distinguished biologist has recently claimed, on the basis of anatomical and physiological studies, that there is a much closer similarity between man and the young, rapidly developing anthropoid than there is between man and the adult ape. According to this, we may be looked upon as arrested or maladjusted apes; while the apes, passing through this stage, go on to adultness, where they cease to struggle for the things they cannot achieve and arrive at reasonable contentment. This is in keeping with Goethe's view that man is a permanent adolescent.

However this may be, it is likely from evidence that, somewhere in the legendary past of louse history, an offspring of a free-living form not unlike our book louse found that life could be infinitely simplified if, instead of having to grub for food in straw, under tree bark, in moss or lichen, in decaying cereals and vegetables, it could attach itself to some food-supplying host, and sit tight. It is one of the few instances in which nature seems extremely logical in its processes. The louse sacrifices a liberty that signifies chiefly the necessity for hard work, the uncertainty of food and shelter, and exposure to dangers from birds, lizards, and frogs; loses the fun of having wings, perhaps; but achieves, instead, a secure and effortless existence on a living island of plenty. In a manner, therefore, by adapting itself to parasitism, the louse has attained the ideal of bourgeois civilization, though its methods are more direct than those of business or banking, and its source of nourishment is not its own species.

Thus, at any rate, arose the parasitic lice, — first, perhaps, the biting ones, the Mallophaga, — and there developed, showing the infinite elasticity of nature: —

The chicken louse
Trinoton, the goose louse
The slender duck louse

The pigeon louse
The turkey louse
The biting guinea-pig louse
Trichodecter, the horse louse

to mention only a few. Out of these, or parallel with them, came the animalcules with which we are chiefly concerned. Not content with a diet of feathers, fur, and dandruff, these varieties — cast off by a kind Providence upon thin-skinned, warm-blooded animals — discovered by an incomprehensible cleverness (or perhaps by an accidental scratch and an occurrence not unlike the discovery of roast pig by the Chinese) that under their feet ran an infinite supply of rich red food. They developed boring and sucking structures, and thus arose: —

The hog louse
The dog louse
Polyplax, the rat louse
The foot louse of the sheep
The cat louse
The short-nosed ox louse
The monkey louse
Our own *pediculi* — the head
  louse and the body louse
  of man

2

It is with the last two that we are chiefly concerned, and they are so closely related that, even now, by an occasional *mésalliance* resulting from the meetings of young people about the neck band, a body louse may go native and interbreed with a head louse. The crab louse we may neglect. He is probably of distinct generic origin and a creature that merits neither respect nor sympathy; not even terror.

Although the human head louse first came into the hair of primitive savages from fur-bearing animals, even in this respect the give-and-take does not appear to have been entirely one-sided. Ewing suggests that the Ateles monkeys may have received their lice from natives; and the similarity between the various monkey lice and those of man is so close that they can interchangeably feed on one or the other host without harm. We have ourselves fed two hundred Arabian head lice on an East Indian monkey for weeks at a time, with relatively low mortality. Such interchange of hosts is not usually possible. A louse fed on a foreign host, in most cases, suffers a probably painful and fatal indigestion.

Ewing further suggests that the spider monkeys obtained their lice from man when the latter reached tropical America in his dispersion from the Old World. The fur of the Ateles monkeys is very similar in coarseness and abundance to that of the head of man, and the blood of this monkey is physiologically more nearly like that of man than that of some other monkeys of the New World. These reflections of Ewing are of great importance in connection with our biography, since the question often arises whether typhus was present in America before the conquest of Mexico. If, as Ewing states, the phylogeny of the Ateles-infesting lice has followed that of their hosts, it is likely that the lice have been in America for a long geological period. The genus Ateles or spider monkey — we quote Elliott from Ewing — has a wide area of distribution, extending from South-Central Brazil as far north as the state of Vera Cruz in Mexico, and from the Pacific coast of Ecuador to the Atlantic coast of Brazil. There are two distinct American groups of *pediculi,* according to our authority — one of them confined to man, and one to monkeys. "The foremost infesting man are largely hybrid head lice, the pure strains of which were originally found on white, black, red, and yellow races living in their original geographical ranges. The monkey-infesting lice of America, so far as known, fall into distinct species according to the hosts they infest, thus indicating to a certain degree at least a parallel phylogeny for host and parasite. If the monkey hosts procured their lice from man, it was not from recent man, but from man that lived tens of thousands of years ago — long enough to allow species differentiation."

Once established on the head of a savage, the louse passed from race to race, acquiring slight changes of form and feature in the process, so that to-day it would seem that we can deduce some information as to human racial relationship from the characteristics of the lice found in different parts of the world. The *Pediculus humanus nigritarum,* or head louse of the African Negro, is slightly different from the head louse found on European and modern American heads. The latter appear to be hybrids, with a strong strain of the nigritarum. The *Pediculus humanus americanus,* found on the prehistoric scalps of American Indian mummies, is again different, and this ancient parasite has been taken from the scalps of living

Indians, together with the European head louse — one among the many acquisitions of civilization.

Our eminent authority, Ewing, studying large series of lice from living Americans, has observed that there was no correlation between louse types and racial types of the human hosts. It appears that America, the melting pot of human races, has also become the melting pot of lice. Ewing became convinced that he was dealing in the American race largely with hybrids of different racial types, and this conviction was strengthened by the relatively recent discovery by Bacot that the head lice of man would intermarry with the body lice and give fertile progeny. This led Ewing — realizing the futility of obtaining any information concerning original American lice from the examination of the heads of our modern intelligentsia — to search for these insects in the scalps of American mummies. At first his search was in vain, because, although he found nits plentiful upon the scalps of pre-Columbian Peruvian mummies, he found no specimens of mummified adults. Later, however, through Dr. Lutz of the American Museum of Natural History, he secured the scalps or hair samples from twenty prehistoric American Indian mummies. Three of these had not only nits, but lice in all stages of development. It was found that the insects from Peruvian mummies were slightly different from those taken in the Southwestern United States, and that all the lice from prehistoric mummy scalps showed differences from some of the lice obtained from a living Indian. It is probable, according to Ewing, that our living Indians have acquired the Caucasian and the Ethiopian head louse, and now enjoy hybrids between these two and the American types. It might be mentioned, also, that the American mummy type is distinct from Fahrenholz's *Pediculus humanus marginatus,* or Japanese variety.

Shipley tells us that the louse adapts its color to that of the host, so that we have the black louse of Africa, the smoky louse of the Hindu, the yellowish-brown louse of the Japanese, the dark brown one of the North American Indian, the pale brown one of the Eskimo, and the dirty gray one of the European.

Again, though the evidence is vague, this prehistoric American louse has been described as quite similar to the Chinese head louse and to the lice found upon Aleutian Eskimos — another argument for the *Völkerwanderung* across the Bering Straits.

From the several head varieties arose the body louse, when naked man began to wear clothing. Primitive races as a rule have no body lice. Advancing in civilized habits with his host, the louse now began to attach its egg cocoons or nits to the fibres of the clothing instead of to the hairs of the body — thereby gaining a degree of protection from direct attack and a greater motility.

In the development of the head louse into the body louse, there are many very interesting changes of habit. Free lice are not often found on the skin. The insects remain in the underclothing in contact with the body, except when feeding, and even at such times they may remain attached by the legs to fibres of the cloth. Soon after conception, the mother louse begins to lay eggs, at the rate of five or more a day, and this is kept up for about thirty days. The eggs are then attached to the fibres of the clothing by a sort of cement substance which forms the nit. Hatching occurs at varying periods, according to the temperature. At normal temperature of the human body, hatching may occur in a week, but if repeatedly exposed to cold or kept at a lower temperature, this process may be delayed for over a month. In getting out of its egg, the young nymph shows extraordinary enterprise. First, it forces open the little lid, or operculum. This gives it the first fascinating glimpse of freedom; but the hole is too small to permit escape. With great ingenuity, the little animal begins to swallow air from in front and eject it from behind, gradually increasing the pressure until eventually it pops out into the great world. It is then a finished little louse, a perfect image of its parents; but if not fed, it dies within a day or two. If properly taken care of, it moults, and in from four days to a week goes into what is spoken of as the second nymph stage, and from that by a similar process into a third nymphal stage, throughout this period enjoying all the privileges of louse existence except the sexual one. It does not become a sexually mature louse until two or three weeks after emerging from the egg. But then . . . Oh, boy!

CHAPTER X

More about the louse: the need for this chapter will be apparent to those who have entered into the spirit of this biography

ALTHOUGH we are aware of the desirability of making progress toward our true objective, the discussion of typhus fever, we may be forgiven if — being on the subject of lice — we devote a few additional pages to these much misunderstood insects. In the study of animal evolution, there seems to have been an almost complete neglect of social forces which, if we study Fabre, Maeterlinck, Wheeler, and others less eminent, appear to play extraordinary rôles in the organization of insect life particularly. The admirably efficient feudal matriarchy of the beehive seems quite superior to any comparable achievement in general contentment developed by man. And the communistic organization of the termites, as described by Professor Wheeler, appears to represent the ultimate perfection of modern Russian aspirations — more perfectly conceived than man seems capable of conceiving them. Yet, in the so-called lower ranges of animal life, we attribute to "instinct" or evolutionary forces the results which men struggle toward with what they call "intelligence." It is at least reasonable to suppose that alterations in human society and government are equally subject to external forces,[1] though man's greater restlessness brings them about with greater speed. In the preceding chapter, we mentioned the possibility that the parasitism developed by the louse was due to the impulse of a bourgeois desire for easy living on the part of the individuals carried by chance to a location where food was simply obtained and life was secure. It is equally possible that there may have been, among these colonists on an abundant soil, a growing conviction that all lice were born equal, that liberty and equality and fraternity should govern society, and that, in this way, the discouragement of wings, of independence, of adventurousness, may have led to a stabilization at the lowest level of louse capacities. However this may be, the louse — like man — has, for one reason or another, failed to develop the highly complex civilization of the bee or the ant. Such development has perhaps been unnecessary because of the infinite and ever-renewed supply of abundant territories for exploration. He lives, blissfully irresponsible, like the Polynesians before the advent of Captain Cook, roaming on the land of plenty, where nature provides warmth, shelter, the odors he loves best, copses for love, and secure undergrowth to which his chosen mate can attach her nest. Under his feet is an inexhaustible supply of the food he prefers, and he has but to sink his hollow stylet into a tender skin to procure his two or three daily meals, with much less trouble than it takes the aborigines to knock a coconut off a tree. In his unrestrained simplicity, he is much like Rousseau's noble savage, — so abhorrent to Mr. Bab-

[1] Professor L. J. Henderson's Seminar on *Pareto* would undoubtedly prove of invaluable assistance in expanding this idea.

bitt, — leading a physically and emotionally unrestricted life.[2] If, with Mr. Babbitt, we deplore this, we cannot — we regret to say — look forward to any changes for the better in the near future. With us, a spiritual deepening is imminent, with the complete exploitation of our continent and the exhaustion of those easy pickings which, for two hundred years, have allowed us to remain, like the louse, undisciplined.[3] But the louse seems indefinitely committed to the materialistic existence, as long as lousy people exist. Each newborn child is a possible virgin continent, which will keep the louse a pioneer — ever deaf to the exhortations of his Van Wyck Brookses and Mumfords better to "evaluate his values."

As far as we can ascertain, since man has existed the

[2] In one important respect, this accusation of Rousseauism is not entirely just to the louse. Though in his other appetites leading an apparently effortless and licentious existence, his sexual arrangements are uniquely wise. Nature has provided that the nymph — that is, what may be called the high-school or flapper age of the louse — is not yet possessed of sexual organs. These do not appear until the fully adult form develops, and reproduction is thus postponed until a responsible age is reached. Adolescent Bohemianism, "living oneself out," "self-expression," and so forth, never get beyond the D. H. Lawrence stage among the younger set. How much physical hardship and moral confusion could be avoided if a similar arrangement among us could postpone sexual maturity until stimulated by an internal secretion from the fully established intellectual and moral convolutions of the brain! The loss of copy this would entail for Theodore Dreiser, William Faulkner, Ernest Hemingway, and others would be amply compensated for by gains in other directions.

[3] Had the Pacific Ocean extended to the west bank of the Mississippi, we should probably have developed, by this time, what is so ardently wished for by our younger critics — a distinctive American culture. With us, the latent seeds planted at Concord a hundred years ago may be expected to burst into flower when the vitality of our race is driven inward by the failure of external resources for material exploitation.

louse has been his inseparable companion. Unlike other parasites, he never leaves his host, except as the consequence of accident or disaster. When he is cast out, or when his host perishes, he is doomed unless he can promptly find another. This fact has led many religiously inclined louse scholars to speculate upon the problem of whether Adam and Eve were lousy. Cowan quotes a writer in the *Gentleman's Magazine* for 1746 as saying, in regard to this fascinating question: "We can hardly suppose that it [the louse] was quartered on Adam and his lady — the neatest pair (if we believe John Milton) that ever joyned hands. And yet, as it disdained to graze the fields or lick the dust for sustenance, where else could it have had its subsistence?" The question can never be settled. We *do* know, however, — as we have elsewhere noted, — that lice are present on the most ancient mummies from many parts of the world, and that these insects were described by early travelers on all savage races encountered by them. Cowan, in his *Curious Facts in the History of Insects*, quotes Wanley's story of the eating of lice by the Budini, a people of Scythia, and the same habit — still prevalent among monkeys — is recorded of the Hottentots and the American Indians. By some of these peoples, as well as by the mediæval English, the practice was supposed to have medicinal value — particularly against the jaundice. In the same extraordinary book, we find citations from *Purchas's Pilgrims* concerning the strange habits of the natives of Malabar, who, "if Lice doe much annoy" them, call upon certain religious and holy men who "will take upon them all those Lice which the others can find and put them upon their [own?] head, there to nourish them" — an act of benevolent self-sacrifice which alone should have served to canonize them.

Pertinent to the now highly probable assumption concerning the prevalence of typhus among the Aztecs before the advent of Cortez is the tale cited from Torquemada. "During the abode of Montezuma among the Spaniards, in the palace of his father, Alonzo de Ojeda one day espied . . . a number of small bags, tied up. He imagined at first that they were filled with gold dust, but on opening one of them what was his astonishment to find it quite full of Lice!" Ojeda spoke of this to Cortez, who then asked Marina and Anguilar for an explanation. He was told that the Mexicans had such a sense of duty to pay tribute to their ruler that the poorest, if they possessed nothing else to offer, daily cleaned their bodies and saved the lice. And when they had enough to fill a bag, they laid it at the feet of their king. Weizl informs us that, when sojourning for a short time among the natives of Northern Siberia, young women who visited his hut sportively threw lice at him. On inquiry concerning this disconcerting procedure, he was embarrassed by learning that this was the customary manner of indicating love, and a notice of serious intentions. A sort of "My louse is thy louse" ceremony.

It is not necessary, however, to confine ourselves to the primitive or ancient races to illustrate the important and intimate rôle played by lice in the social life of the human race. Among the unfortunates of our own day, these little creatures are still sufficiently prevalent in the most civilized communities, although, in places as decadent as Boston is said to be by Upton Sinclair, it is often difficult to find a needed supply of the insects, unless one knows one's way about. In our experience, on one occasion — when a supply of uninfected lice was needed immediately for feeding on a suspected case of typhus fever — it became necessary, by appeal to the scientific enthusiasm of a municipal police captain, to place under temporary arrest a colored gentleman who was the only individual easily discovered who was in possession of the coveted insects. It is needless to add that he was, of course, immediately released — after generously supplying us from his ample store.

Yet, as everyone who has really been to war knows, let the water supply fail, or soap become scarce, or a change of clothing be delayed — it takes no time at all before the louse comes back to its own. It was not so long ago, indeed, that its prevalence extended to the highest orders of society, and was accepted as an inevitable part of existence — like baptism, or the smallpox.

Lice have even been important in politics. Cowan tells the story of the custom prevailing in Hurdenburg in Sweden, where in the Middle Ages a mayor was elected in the following manner. The persons eligible sat around a table, with their heads bowed forward, allowing their beards to rest on the table. A louse was then put in the middle of the table. The one into whose beard the louse first adventured was the mayor for the ensuing year.

The manner of living throughout the Middle Ages made general lousiness inevitable. In England, in the twelfth and thirteenth centuries, the houses of the poor were mere hovels, often with only a hole in the roof to let out the smoke of the central fire; and in cold weather the families were huddled together at night without changing the simple garments — usually a single shift — which they wore in the daytime. Washing was practically out of the question, and the better classes — not very much more comfortable in their badly heated domiciles — wore a great many clothes, which they rarely changed. MacArthur's story of Thomas à Becket's funeral illustrates this: —

The Archbishop was murdered in Canterbury Cathedral on the evening of the twenty-ninth of December. The body lay in the Cathedral all night, and was prepared for burial on the following day. The Archbishop was dressed in an extraordinary collection of clothes. He had on a large brown mantle; under it, a white surplice; below that, a lamb's-wool coat; then another woolen coat; and a third woolen coat below this; under this, there was the black, cowled robe of the Benedictine Order; under this, a shirt; and next to the body a curious haircloth, covered with linen. As the body grew cold, the vermin that were living in this multiple covering started to crawl out, and, as MacArthur quotes the chronicler: "The vermin boiled over like water in a simmering cauldron, and the onlookers burst into alternate weeping and laughter."

The habit of shaving the head and wearing a wig was

no doubt in part due to the effort to hold down vermin. Gentlemen and ladies all over Europe resorted to this, but the wigs they wore were often full of nits. Pepys speaks of this in several places, complaining about a new wig he had bought which was full of nits. "Thence to Westminster to my barber's; to have my Periwigg he lately made me cleansed of its nits, which vexed me cruelly that he should put such a thing into my hands."

Even in the highest society, the questions of lice and scratching were serious problems; and the education of children, even in the highest circles, included a training of the young in relation to their vermin. Reboux, speaking of the education of a princess of France in the middle of the seventeenth century, says: "One had carefully taught the young princess that it was bad manners to scratch when one did it by habit and not by necessity, and that it was improper to take lice or fleas or other vermin by the neck to kill them in company, except in the most intimate circles."

He tells another story illustrative of the universal lousiness even of the aristocracy. The young Comte de Guiche had made himself unpopular with the King by casting amorous eyes upon Madame, the King's sister-in-law. He sent the Comte's father to announce banishment to the son. The latter was not yet out of bed when his father arrived. As the old Marshal stood in front of the bed, a louse crawled out from under his perruque, began to crawl along the deep furrows on the old man's forehead, skirted the edges of the little thickets made by the eyebrows, and crawled back under the hair of the wig.

The entire lecture was missed, while the Comte de Guiche was watching the adventures of the insect.

Even long into the eighteenth century, lice were regarded as necessities. Bacteriologists for a generation have wondered whether the presence of colon bacilli in the intestines might not, because of their universal occurrence, have some physiological purpose. For similar reasons, as wise a man as Linnæus suggested that children were protected by their lice from a number of diseases.

In the story of George Washington by Rupert Hughes, we find the following paragraph on "Rules of Civility," copied by Washington in his fourteenth year: "Kill no vermin, as Fleas, lice, tics, etc. in the sight of others, if you See any filth or thick Spittle, put your foot Dexteriously upon it; if it be upon the Cloths of your Companions, put it off privately, and if it be upon your own Cloths, return thanks to him who puts it off."

Since Colonial days, these things have changed. The louse has been banished completely from fashionable society, and even though — among our middle classes — there may not be a motor car in every garage, there is almost invariably a bathtub in every cottage and flat. And more and more, the habit of keeping the coal in the bathtub is disappearing. The louse is confined, in consequence, to the increasingly diminishing populations of civilized countries who live in distress and great poverty. But there are still many of these with us, and there are regions of the earth where life is still primitive, where bathtubs remain luxuries and bathing amounts to counterrevolution. The louse will never be completely exterminated, and there will always be occasions when it will spread widely to large sections of even the most sanitated populations.

And as long as it exists, the possibility of typhus epidemics remains.

CHAPTER XI

Much about rats — a little about mice

I

It is now quite well established that the subject of our biography is, in some phases of its adventurous existence, closely associated with rats. Since it is our purpose to write a well-balanced account, undistorted by exaggerated emphasis or by omissions, it becomes necessary to give some attention to these rodents, which play a rôle as important in the history of mankind as the other hosts of typhus. In dealing with rats, we must consider as well, though in a minor way, their smaller brethren, the mice — not only because what rats can do, mice may also accomplish, but because typhus virus can be kept comfortable and alive in some mice, which means that they also become subjects for further epidemiological study. A close relative of our own typhus, the Tsutsugamushi fever of Japan, is actually conveyed, from field mice to man, by the harvest mite.[1]

In regard to the association of rats with typhus, the known facts are still, in a degree, rudimentary. All that we know definitely is that the virus of the typhus fever of the New World has been found in the rat flea and in the brains of rats trapped in an epidemic focus in Mexico City. The disease, in the places mentioned, may be con-

[1] For the following classification of the rodents, we are indebted to Professor Paul A. Moody (excerpt from Rodent Classification, based on

veyed from the infected rat to man by the flea. We know, also, that rats in the Mediterranean basin are similarly infected. Studies made within the last few years seem to indicate that the virus of the Mexican-American type of typhus fever, as well as of the endemic variety in the Mediterranean basin, is highly adapted to rodents and is carried in these animals — rats — during the intervals between human epidemics; transmitted from rat to rat by the rat louse (polyplax) and the rat flea (Xenopsylla), and, on suitable occasions, to man from the rat by the rat flea. For this reason, Nicolle speaks of this as the "murine" virus. The virus obtained from cases occurring in the historic Eastern European typhus foci and in Africa is less virulent for rodents, and there is reason to believe, from observations too technical to be here described, that this virus has been propagated for centuries, not only in rats,

List of North American Recent Mammals, 1923, by Gerrit S. Miller, Jr., Bull. 128, U. S. National Museum): —
ORDER RODENTIA:
  Superfamily Muridæ:
    Family Cricetidæ (New World):
      Subfamily Cricetinæ:
        Genus *Peromyscus*, deer mice
        Genus *Sigmodon*, cotton rats
      Subfamily Microtinæ:
        Genus *Microtus*, meadow mice
        (Subfamily also includes lemmings and muskrats, among others)
    Family Muridæ (New World):
      Subfamily Murinæ:
        Genus *Micromys*, European harvest mice
        Genus *Rattus*:
          Rattus rattus, black rat, formerly called *Epimys rattus*
          Rattus norvegicus, Norway or house rat
        Genus *Mus*, house mice

but in human carriers. With infected human beings, the European infection has been imported to America as what is known as "Brill's disease," so that we now enjoy both varieties on this continent. It is more than likely that in both cases the virus comes from a common stock which originally infected rodents. Hence our preoccupation with these animals. They become objects of interest in tracing the epidemiology not only of typhus, but also of plague — these two calamities sharing with human ferocity the greatest responsibility for wholesale sorrow, suffering, and death throughout the ages.

It is a curious fact that long before there could have been any knowledge concerning the dangerous character of rodents as carriers of disease, mankind dreaded and pursued these animals. Sticker has collected a great many references to this subject from ancient and mediæval literature, and has found much evidence in the folklore of mediæval Europe which points to the vague recognition of some connection between plague and rats. In ancient Palestine, the Jews considered all seven mouse varieties (akbar) unclean, and as unsuited for human nourishment as were pigs. The worshipers of Zoroaster hated water rats, and believed that the killing of rats was a service to God. It is also significant that Apollo Smintheus, the god who was supposed to protect against disease, was also spoken of as the killer of mice, and Saint Gertrude was besought by the bishops of the early Catholic Church to protect against plague and mice. The year 1498, Sticker tells us, was a severe plague year in Germany, and there were so many rats in Frankfurt that an attendant was sta-

tioned for several hours each day on a bridge in the town
and directed to pay a pfennig for every rat brought in.
The attendant cut off the tail of the rat — probably as a
primitive method of accounting — and threw the bodies
into the river. Heine, according to Sticker, speaks of a tax
levied on the Jews of Frankfurt in the fifteenth century,
which consisted of the annual delivery of five thousand
rat tails. Folklore originating in a number of different
parts of Europe during the great plague epidemics mentions cats and dogs, the hereditary enemies of rats and
mice, as guardians against the plague.

Most scholars agree that there is no reliable mention of
rats — as such — in classical literature. The Greeks had
the word μῦς. Herodotus mentions the field mouse —
μῦς ἀρουραῖος. The expression μῦς ἐν πίττῃ (mouse in a
pickle jar) meant "to be in a bad hole or scrape." The
Greeks also knew ὕραξ, — the later Roman "Sorex," —
which, though not a rodent at all (the shrewmouse),
looked enough like one to get into the literature with the
mouse. Our learned friend Professor Rand tells us of
a story quoted by Keller (*Die Antike Thierwelt*) about
Heliogabalus, who "staged a fight between ten thousand
mice, one thousand shrewmice and one thousand weasels."
Needless to relate, the shrewmice "polished off" the mice,
and the weasels got both of them.[2]

---

[2] In connection with this story about Heliogabalus, it is particularly
strange that rats were not included in his curious amusement. According to Hamilton and Hinton, the rat was "undoubtedly present in
the East before the time of the Crusaders, and was firmly established
in Europe shortly after 1095." Heliogabalus, whose real name was
Varius Avitus, a native of Emesa, was taken from Rome to his birth-

The Romans knew the mouse well. It was recognized as a pest, and *musculus* (little mouse) was even used as a term of endearment by Martian. The word root (*muishi*, Persian; *musa, musi*, Hindu; *musiko*, Pali) indicates the world-wide ancient knowledge of mice.

There is, however, no specific early differentiation between mice and rats, and authorities seem to agree quite generally that nothing in the references to mice, at least among the Greeks and Romans, justifies the assumption that rats may have been referred to. Yet, in view of the probable ancient prevalence of rats in Eastern countries, and the close communications by sea between the Greeks and the Mediterranean coastal cities, as well as the regular grain traffic between Egypt and Rome, it is difficult to credit the complete absence of rats from the European littoral throughout antiquity.

In regard to mice and rats in the Near East, Herodotus tells us of Libya that "in this country there are three

place, Emesa, after the murder of Caracala. There he became high priest of the Syrian sun god, Elagabalus, whose name he assumed. On his return to Rome as Emperor in 219 A.D., he affronted the community by, among other things, the "horseplay and childish practical joking" of which the mouse battle is an example. In attempting to determine whether rats reached Rome at or about the time when free communication between Rome and the Levant was habitual, in the centuries following the year one, it is of interest to note that when the true black rat surely arrived in Italy after the Crusades, it was known as "Sorco," from "Sorex" — which justifies the suspicion that the Sorex of Heliogabalus might have been a rat. This is further encouraged by the thought that rats might more easily have carried the victory over the mice than true shrewmice, which are insectivorous and not very large or powerful. In later literature, according to Hamilton and Hinton, the black rat has figured as "Sorex" (Hoefnagel-Archetypa, 1592), and is referred to as *Mus major seu Sorex* in Merrett (Pinax, 1667).

kinds of mice. One is called the 'two-legged' mouse; another the 'Zegeris' [a word that means a hill — possibly a sort of prairie dog]; a third, the 'prickly' mouse." Also he recounts that when Sanachrib, King of Arabia and Assyria, marched a great host against Egypt, on the night before the battle "there swarmed upon them mice of the fields, and ate up their quivers and their bows and the handles of their shields" so that, on the next day, they fled. This sounds much more like rats than like the timid field mouse. However, these things are hardly evidence.[3]

It is quite impossible to make a case for the presence of true rats in Europe proper during classical times, much as this would clarify the epidemiological situation. It is conceivable that the manner of transmission of plague and typhus may have undergone modification since the Peloponnesian Wars by changed adaptations to hosts, both insect and rodent. But it would seem much more likely that the zoölogical differentiations between rodents so similar and closely related as mice and rats were inaccurate in ancient records, and that rats may have existed — though undomesticated. This would give us a wider latitude for speculation regarding the nature of epidemics, which, to be sure, were rarely, under the circumstances of ancient life, as widespread or deadly as they became with the later concentrations of population and of urban habits. At any rate, if rats had been present in those times in anything like the numbers in which they are found to-day, we should probably have reliable records. It may well be

[3] The same story is found in Josephus.

that the frugality of well-run households, like that of Penelope, gave little encouragement to house rats to become parasitic on man to the extent to which they have since.

All this is conjecture. According to the wisest students of the subject, there is no certain knowledge of rats in Europe, within historic periods, until shortly after the Crusades. In prehistoric days they certainly existed there — but later disappeared. Fossil remains of rats have been found in the Pliocene period of Lombardy (the Mastodon period of Europe) and in the later Pleistocene of Crete. They were present during the glacial period with the lake dwellers, whom they pestered in Mecklenburg and Western Germany. From that time on, there were either few or no rats until thousands of years later.

In regard to the reappearance of rats in Europe, our industrious colleagues, the zoölogists, have gathered an immense amount of information, much of which has been interestingly summarized by Barrett-Hamilton and Hinton in their *History of British Mammals,* and by Donaldson in his *Memoir on the Rat.* Before we proceed to this subject, however, it will be profitable to consider the striking analogy between rats and men. More than any other species of animal, the rat and mouse have become dependent on man, and in so doing they have developed characteristics which are amazingly human.

In the first place, like man, the rat has become practically omnivorous. It eats anything that lets it and — like man — devours its own kind, under stress. It breeds at all seasons and — again like man — it is most amorous in the springtime.[4] It hybridizes easily and, judging by the strained relationship between the black and the brown rat, develops social or racial prejudices against this practice. The sex proportions are like those among us. Inbreeding takes place readily. The males are larger, the females fatter. It adapts itself to all kinds of climates. It makes ferocious war upon its own kind, but has not, as yet, become nationalized. So far, it has still stuck to tribal wars — like man before nations were invented. If it continues to ape man as heretofore, we may, in a few centuries, have French rats eating German ones, or Nazi rats attacking Communist or Jewish rats; however, such a degree of civilization is probably not within the capacities of any mere animal. Also — like man — the rat is individualistic until it needs help. That is, it fights bravely alone against weaker rivals, for food or for love; but it knows

---

[4] On first sight, the fertility of rats would seem far to outstrip that of man; for rats reach adolescence when a little more than half grown, and produce one or two litters a year, averaging from five to ten in number. The difference from man, however, is not so striking if one remembers Donaldson's calculation that one rat year equals thirty years for man, and makes the comparison with human society of former years — in savage communities, or before the humane and sane practice of birth control had begun to weaken the inhibitions of religion in such matters. Many examples not too unlike conditions among rats could be cited — such as, for instance, the story of Samuel Wesley, father of John, which we take from a review by J. C. Minot of Laver's biography of Wesley. Samuel had fourteen children with his good Sukey before 1701, when he left her because she refused to pray for William III as the lawful King of England. On the accession of Queen Anne, he was reconciled and bestowed five more children upon the fortunate woman. The oldest of these pledges of reconciliation was the immortal John Wesley.

how to organize armies and fight in hordes when necessary.

Donaldson, basing his calculations mainly on stages in
the development of the nervous system, reckons three
years of a rat life as ninety years for man. By this scale, the
rat reaches puberty at about sixteen, and arrives at the
menopause at the equivalent of forty-five. In following
man about all over the earth, the rat has — more than any
other living creature except man — been able to adapt
itself to any conditions of seasonal changes or climate.

2

The first rat to arrive in Europe was *Mus rattus* — the
black rat, house rat, or ship rat. It may have wandered in
between 400 and 1100 A.D., with the hordes that swept into
Europe from the East in that period of great unrest —
the *Völkerwanderung*. It may not have arrived until
somewhat later, when the first Crusaders returned. It is
not mentioned in the Epinal Glossary of 700 A.D., but
may have been meant by the word "raet" in the English
Archbishop Ælfric's Vocabulary of 1000 A.D. But the authorities from whom we cite this call attention to the fact
that the word "rata" was the Provençal for the domestic
mouse of that time, and the word may have been introduced into England.[5] Hamilton and Hinton say that the

[5] Rats and mice belong to the same genus, and the closeness of the
relationship is attested by the experiment of Ivanoff, who artificially
inseminated a white mouse with the sperm of a white rat, and obtained
two hybrids after a pregnancy of twenty-seven days. Mice may have
developed out of rats under circumstances which made it less desirable
to be large and ferocious than to be able to get into a smaller hole —

first clear differentiation between rats and mice is found in the writings of Giraldus Cambrensis (1147–1223). After that date, it is referred to frequently.

As to the Eastern origin of the black rat, there seems to be no difference of opinion among authorities, though there is much uncertainty about the exact part of the Orient from which it came. De L'Isle believes that the *Mus alexandrinus* represents the source stock of the European *Mus rattus*. This — the Alexandrine rat — did not, according to him, become parasitic on human society until the seventh century — living before this time a wild existence, possibly in the Arabian deserts, a fact which would account for its failure to migrate into classical Europe with trade, and, in the early Middle Ages, with Saracen invasions. By the time of the Crusaders, it had begun to domesticate and consequently to follow human travel. Being a climber and therefore a ship rat, it spread rapidly to Mediterranean ports, where, according to Hamilton and Hinton, its arrival by sea is witnessed to by the name ποντικος applied to it by the modern Greeks; "pantagena" by the Venetians. The Genoese mistook it for a mole, calling it "Salpa," another point of evidence that it may have been new to them.

From the time of its arrival, the rat spread across Europe with a speed superior even to that of the white man in the Americas. Before the end of the thirteenth century, it had become a pest. The legend of the *Rattenfänger von Hameln*, who piped the children into the

the advantages of which may be appreciated by those of us who have lived in the world during the post-war years.

hollow Koppenberg because the town refused his pay for
piping the rats into the Weser, is placed at or about 1284.
By this time, the rat had penetrated into England. It had
reached Ireland some time before this, where it was the
"foreign" or "French" mouse, "ean francach." Our authorities tell us that in Ireland, even until very recent
times, everything foreign was called "francach," or French.
A little later, the rat was in Denmark, Norway, and the
adjacent islands. By Shakespeare's time, the black rat was
so formidable a nuisance that days of prayer for protection
against its ravages were set aside, and rat catchers (see
Romeo and Juliet, Act III) were important officials,
probably calling themselves, as they would to-day, scientists or artists (or "rattors" — *cf.* "realtors" and "morti-
cians").

For twice as long as the Vandals had their day in North
Africa, or the Saracens in Spain, or the Normans in Italy,
the black rats had their own way in Europe. Their reign
covered the periods of the devastating epidemics of
plague that swept through the battle areas of the Thirty
Years' War and the later ones of the seventeenth century.
And during the centuries of its supremacy there occurred
the most destructive typhus epidemics, accompanying
wars and famines, that have occurred up to our own time.
Whether the black rats of mediæval Europe played a
rôle in these remains uncertain. That they played the
leading part in the plague epidemics of this time seems
beyond question.

But just as the established civilizations of Northern Europe were swept aside by the mass invasions of barbarians

from the East, so the established hegemony of the black rat was eventually wiped out with the incursion of the hordes of the brown rat, or *Mus decumanus* — the ferocious, short-nosed, and short-tailed Asiatic that swept across the Continent in the early eighteenth century; until at the present time, the slender-nosed, long-tailed, climbing *Mus rattus* has been all but exterminated in its former strongholds, and continues to thrive only in relatively small groups along the littoral, in seaports, on islands, or in countries like South America and other tropical regions where it is not confined to parasitic life in competition with its larger and more barbaric rival, or where the brown *conquistadores* have not yet arrived. It maintains its former superiority only on ships, where, because of its greater ability in climbing, it can still hold its own.[6]

The brown rat, too, came from the East. It is now known as the "common" rat and, because of a mistaken notion of its origin, as *Mus norvegicus*. Its true origin, according to Hamilton and Hinton, is probably Chinese Mongolia or the region east of Lake Baikal, in both of which places forms resembling it have been found indigenous. The same writers quote Blasius, who believes that the ancients about the Caspian Sea may have known this rat. Claudius Ælianus, a Roman rhetorician of the second century, in his *De Animalium Natura*, speaks of "little less than Ichneumons, making periodical raids in infinite numbers" in the countries along the Caspian, "swimming over rivers holding each other's tails." This

---

[6] In a recent rat survey of Boston, black rats were found in only a single small and circumscribed area, close to the docks.

may or may not be so; but it seems certain that this rat was not known in Western Europe until the eighteenth century.

Pallas (1831), in his *Zoögraphica Rosso-Asiatica,* records that in 1727 — a mouse year — great masses of these rats swam across the Volga after an earthquake. They invaded Astrakhan, and thence rapidly spread westward. They reached England, probably by ship, in 1728, and were unjustly called the "Hanoverian rat" because of the unpopularity of the House of Hanover, though probably they had not arrived in Germany at that time. They were seen in Prussia in 1750, and were common by 1780. This rat was unknown to Buffon in 1753 and to Linnæus in 1758 — but both of these gentlemen were already "famous" scientists at this time, and most likely occupied in attending committee meetings. The brown rat arrived in Norway in 1762, a little later in Spain, and in Scotland about 1770. By 1775 it had come to America from England. It appears to have had a hard time only in countries where the population is what is spoken of as "thrifty." In Scotland, it took from 1776 to 1834 to get from Selkirk to Morayshire; it did not dare enter Switzerland until 1869, and has never done very well among the Switzers. It spread slowly across our continent, owing to deserts, rivers, and long distances between "hand-outs." Consequently, it did not arrive in California until shortly after 1851. Now that it is there, it thrives in that wonderful climate as hardly elsewhere. At the present time the rat has spread across the North American Continent from Panama to Alaska, has penetrated to all the less tropical parts of South America, to the South Sea Islands, to New Zealand, and to Australia. In fact, it has conquered the world. Only the extreme cold of Greenland does not seem to attract it. Unlike the Eskimo, it has had the good sense, whenever introduced to the arctic regions, to wander southward at the first opportunity.

Wherever it has gone, it has driven out the black rat and all rival rodents that compete with it. From the point of view of all other living creatures, the rat is an unmitigated nuisance and pest. There is nothing that can be said in its favor.[7] It can live anywhere and eat anything. It burrows for itself when it has to, but, when it can, it takes over the habitations of other animals, such as rabbits, and kills them and their young. It climbs and it swims.

It carries diseases of man and animals — plague, typhus, trichinella spiralis, rat-bite fever, infectious jaundice, possibly Trench fever, probably foot-and-mouth disease and

[7] Of course, rats might form a cheap source of food. They have been eaten without harm under stress — at the siege of Paris in 1871, and before that by the French garrison at Malta in 1798, where, according to Lantz, food was so scarce that a rat carcass brought a high price. The same writer states that Dr. Kane of the arctic ship *Advance* ate rats through the winter, and avoided scurvy — from which his more fastidious companions all suffered. For the following story we cannot vouch. It is related to us that a learned specialist on rodents was lecturing, some years ago, in one of the more distinguished university centres in the United States. After the lecture, he was taken to a restaurant famous for its terrapin. He enjoyed his meal and praised the quality of the *pièce de résistance*, but recognized the bones on his plate as those of rats. He is said later to have visited the albino rattery where the "terrapin" was bred. The matter might be looked into as a commercial possibility. Robert Southey once suggested that the first requisite to successful rat eradication was to make them a table delicacy.

a form of equine "influenza." Its destructiveness is almost unlimited. Lantz, of the United States Department of Agriculture, has made some approximate estimates of this, as follows (we abbreviate): —

Rats destroy cultivated grain as seeds, sprouts, or after ripening.

They eat Indian corn, both during growth and in the cribs, and have been known to get away with half of the crop. A single rat can eat from forty to fifty pounds a year.

They destroy merchandise, both stored and in transit, books, leather, harness, gloves, cloth, fruit, vegetables, peanuts, and so forth.

The rat is the greatest enemy of poultry, killing chicks, young turkeys, ducks, pigeons; also eating enormous numbers of eggs.

Rats destroy wild birds, ducks, woodcocks, and song birds.

They attack bulbs, seeds, and young plants or flowers.

They cause enormous damage to buildings, by gnawing wood, pipes, walls, and foundations.

Hagenbeck had to kill three elephants because the rats had gnawed their feet. Rats have killed young lambs and gnawed holes in the bellies of fat swine.

They have gnawed holes in dams and started floods; they have started fires by gnawing matches; they have bitten holes in mail sacks and eaten the mail; they have actually caused famines in India by wholesale crop destruction in scant years.

They have nibbled at the ears and noses of infants in

their cribs; starving rats once devoured a man who entered a disused coal mine.

3

A rat census is obviously impossible. It is quite certain, however, that they breed more rapidly than they are destroyed in many places in the world. We can appraise the rat population only by the numbers that are killed in organized rat campaigns and by the amount of destruction they cause. In about 1860, Shipley tells us, there was a slaughterhouse for horses on Montfaucon, which it was planned to remove farther away from Paris. The carcasses of horses amounted to sometimes thirty-five a day, and were regularly cleaned up completely by rats in the following night. Dusaussois had the idea of trying to find out how many rats were engaged in this gruesome traffic. He set horse-meat bait in enclosures from which the exit of rats could be prevented, and in the course of the first night killed 2650. By the end of a month, he had killed over 16,000. Shipley estimates that there are about forty million rats in England at one time. In 1881 there was a rat plague in certain districts of India. The crops of the preceding two years were below average and a large part of them had been destroyed by rats. Rewards offered for rat destruction led to a killing of over 12,000,000 rats. Shipley estimates that a single rat does about 7s. 6d. worth of damage in a year, which makes a charge of £15,000,000 upon Great Britain and Ireland. It costs about sixty cents to two dollars a year to feed a rat on grain. Every rat on a farm costs about fifty cents a year. Lantz adds to this that hotel managers estimate five dollars a year as a low estimate of the loss inflicted by a rat. He thinks that in the thickly populated parts of the country an estimate of one rat per acre is not excessive, and that in most of our cities there are as many rats as people. He investigated, in 1909, the approximate total damage by rats in the cities of Washington and Baltimore. From the data he obtained, he calculated the annual damage in the two cities as amounting to $400,000 and $700,000 respectively — which, considering the populations, amounted to an average loss of $1.27 a year per person. On the same basis, the urban population of the United States, at that time 28,000,000 people, sustained an annual direct injury of $35,000,000 a year. In Denmark, the estimated rat cost is about $1.20 a person; in Germany, eighty-five cents a person; in France, a little over a dollar. Add to this the inestimable depreciation of property and the costs of protection.

All this has nothing to do with our main subject, but we were started on rats, and it is just as well to give thought to the problem of what rat extermination for sanitary purposes is likely to mean in other respects.

The tremendous speed with which rats swarmed over the continents of the world can be readily understood if one reads the observations of actual rat migrations made in modern times. The seasonal migration of rats from buildings to the open fields takes place with the coming of the warm weather and the growth of vegetation; and a return to shelter follows with the cold weather. Dr. Lantz tells us that in 1903 hordes of rats migrated over

several counties in Western Illinois, suddenly appearing
when for several years no abnormal numbers had been
seen. An eyewitness stated to Lantz that, as he was returning to his home on a moonlight night, he heard a
rustling in a near-by field, and saw a great army of rats
cross the road in front of him. The army of rats stretched
away as far as he could see in the moonlight. This, to be
sure, was before the Eighteenth Amendment, but there
must have been some fact behind it, since heavy damage
was caused by rats in the entire surrounding country of
farms and villages in the ensuing winter and summer. On
one farm, in the month of April, about 3500 rats were
caught in traps. Lantz himself saw a similar migration in
the valley of the Kansas River, in 1904; and Lantz, being at that time an officer and gentleman of the United
States Agricultural Service, cannot be under the suspicion
that is aroused by accounts of armies of rats seen by moon-
shine. In England a general movement of rats inland
from the coast occurs every October, and this migration
is connected with the closing of the herring season. During the herring catch, rats swarm all over the coast, at-
tracted by the food supply of herring cleaning; when it
is over, they go back to their regular haunts. In South
America, Lantz advises us, rat plagues are periodic in
Paraná, in Brazil, and occur at intervals of about thirty
years. In Chile, the same thing has been observed, at
intervals of fifteen to twenty-five years. Studies of these
migrations have shown that the rat plagues are associated
with the ripening and decay of a dominant species of bamboo in each country. For a year or two, the ripening seed

in the forests supplies a favorite food for the rats. They multiply enormously, and eventually, this food supply failing, they go back to the cultivated areas. A famine was caused in 1878 in the state of Paraná by the wholesale destruction of the corn, rice, and mandioca crops by rats. The invasion of Bermuda by rats in 1615, and their sudden disappearance, are as dramatic as the rise and fall of some of the short-lived Indian empires of Central and South America. Black rats appeared in that year, and within the two following ones increased with alarming rapidity. They devoured fruits, plants, and trees to such an extent that a famine resulted, and a law required every man in the islands to keep twelve traps set. Nothing, however, was of any use, until finally the rats disappeared with a suddenness that makes it almost necessary to assume that they died of a pestilence.

As we have indicated in a preceding paragraph, the natural history of the rat is tragically similar to that of man. Offspring of widely divergent evolutionary directions, men and rats reached present stages of physical development within a few hundred thousand years of each other — since remnants of both are found in the fossils of the glacial period.

Some of the more obvious qualities in which rats resemble men — ferocity, omnivorousness, and adaptability to all climates — have been mentioned above. We have also alluded to the irresponsible fecundity with which both species breed at all seasons of the year with a heedlessness of consequences which subjects them to wholesale disaster on the inevitable, occasional failure of the food supply. In this regard, it is only fair to state — in justice to man — that, as far as we can tell, the rat does this of its own free and stupid gluttony, while man has tradition, piety, and the duty of furnishing cannon fodder to contend with, in addition to his lower instincts. But these are, after all, phenomena of human biology, and man cannot be absolved of responsibility for his stupidities because they are the results of wrong-headedness rather than the consequences of pure instinct — certainly not if they result in identical disasters.

Neither rat nor man has achieved social, commercial, or economic stability. This has been, either perfectly or to some extent, achieved by ants and by bees, by some birds, and by some of the fishes in the sea. Man and the rat are merely, so far, the most successful animals of prey. They are utterly destructive of other forms of life. Neither of them is of the slightest earthly use to any other species of living things. Bacteria nourish plants; plants nourish man and beast. Insects, in their well-organized societies, are destructive of one form of living creature, but helpful to another. Most other animals are content to lead peaceful and adjusted lives, rejoicing in vigor, grateful for this gift of living, and doing the minimum of injury to obtain the things they require. Man and the rat are utterly destructive. All that nature offers is taken for their own purposes, plant or beast.

Gradually these two have spread across the earth, keeping pace with each other and unable to destroy each other, though continually hostile. They have wandered from East to West, driven by their physical needs, and — un-

like any other species of living things — have made war
upon their own kind. The gradual, relentless, progressive extermination of the black rat by the brown has no
parallel in nature so close as that of the similar extermination of one race of man by another. Did the Danes conquer England; or the Normans the Saxon-Danes; or the
Normans the Sicilian-Mohammedans; or the Moors the
Latin-Iberians; or the Franks the Moors; or the Spanish
the Aztecs and the Incas; or the Europeans in general the
simple aborigines of the world by qualities other than
those by which *Mus decumanus* has driven out *Mus
rattus?* In both species, the battle has been pitilessly to the
strong. And the strong have been pitiless. The physically
weak have been driven before the strong — annihilated, or
constrained to the slavery of doing without the bounties
which were provided for all equally. Isolated colonies of
black rats survive, as weaker nations survive until the
stronger ones desire the little they still possess.

The rat has an excuse. As far as we know, it does not
appear to have developed a soul, or that intangible quality of justice, mercy, and reason that psychic evolution
has bestowed upon man. We must not expect too much.
It takes a hundred thousand years to alter the protuber-
ances on a bone, the direction of a muscle; much longer
than this to develop a lung from a gill, or to atrophy a
tail. It is only about twenty-five hundred years since
Plato, Buddha, and Confucius; only two thousand years
since Christ. In the meantime, we have had Homer and
Saint Francis, Copernicus and Galileo; Shakespeare, Pascal, Newton, Goethe, Bach, and Beethoven, and a great

number of lesser men and women of genius who have demonstrated the evolutionary possibilities of the human spirit. If such minds have been rare, and spread thinly over three thousand years, after all, they still represent the sports that indicate the high possibilities of fortunate genetic combinations. And these must inevitably increase if the environment remains at all favorable. If no upward progress in spirit or intelligence seems apparent, let us say, between the best modern minds and that of Aristotle, we must remember that, in terms of evolutionary change, three thousand years are negligible. If, as in the last war and its subsequent imbecilities, mankind returns completely to the rat stage of civilization, this surely shows how very rudimentary an emergence from the Neanderthal our present civilization represents — how easily the thin, spiritual veneer is cracked under any strain that awakens the neolithic beast within. Nevertheless, for perhaps three or five thousand years, the beast has begun to ponder and grope. Isolated achievements have demonstrated of what the mind and spirit are capable when a happy combination of genes occurs under circumstances that permit the favored individual to mature. And the most incomprehensible but hopeful aspect of the matter is the fact that successive generations have always bred an adequate number of individuals sufficiently superior to the brutal mass to keep alive a reverence for these supreme achievements and make them a cumulative heritage. It is more than likely — biologically considered — that by reason of this progressive accumulation of the best that superior specimens of our species have produced, the evolution toward higher things may gain velocity with time, and that in another hundred thousand years the comparison of the race of men with that of rats may be less humiliatingly obvious.

Man and the rat will always be pitted against each other as implacable enemies. And the rat's most potent weapons against mankind have been its perpetual maintenance of the infectious agents of plague and of typhus fever.

CHAPTER XII

We are at last arriving at the point at which we can approach the subject of this biography directly. We consider intimate family relations, immediate ancestors, and gestation of typhus

I

A GREAT deal of that which has gone before was incidental to our scrutiny of the literature of infectious diseases — undertaken for the purpose of ascertaining how early in recorded history typhus fever was recognizably described. The search turned up so many side issues and suggested so much that it amused us to discuss that we wandered from one digression to the next, following our own inquisitive nose and completely forgetting the reader, who, after all, was led — by our introductory chapter — to assume that he was about to read of typhus fever. Apologetically, therefore, and not without some astonishment, we discover that most of our book has run out of the pen, and the purpose for which it was undertaken remains unaccomplished. The temptation of discursiveness is a strong one, and we are, even now, lured by reminiscences of troublous times in epidemic regions of post-war Europe, again postponing typhus, to consider the degree to which pestilence and famine have contributed to the economic and social upheavals of that disturbed continent. Will historians of this period remember that, throughout the struggles which led to the establishment of the Soviet Republic, Russia suffered — in addition to war and armed revolution — from two cholera epidemics, from a famine unequaled since the Thirty Years' War, from typhus, malaria, typhoid, dysentery, tuberculosis, and syphilis to an extent unimaginable except to those who were helpless spectators? Tarassewitch estimated (statistics of accuracy were impossible) that between 1917 and 1923 there were 30,000,000 cases of typhus with 3,000,000 deaths in European Russia alone.

Tarassewitch — what a man he was! We think of him in moments of depression and take courage from his spirit. We remember, as though we had been privileged to dine with a king, the breakfast of cheese and bread and tea which we had at his table. "After all, this is my country," he said. "There are few of us left who have been trained to this work. I am a Russian, and these are my people." He said it like a simple gentleman, whimsically bashful, utterly without dramatization, as though he feared we might think him indulging in heroics. He had innumerable chances of escaping from conditions that deprived him of everything except the opportunity of sharing the sufferings of an unhappy nation. He and others like him, — Zabolotny, Korschun, Barykin, — they knew that they were fighting a rear-guard action, but they stood by, proudly unmindful of insult, humiliation, and penury, because they hoped to be able to hold together the remnants of their thinning ranks for services which no others could render, and which they knew that Russia would need, whatever her political destinies. Standing before me in his house in Moscow, in meagre linen blouse and trousers, with sandals instead of shoes, there was in him a fine arrogance and gallantry as he said these things. There were others like him. Most of them are dead and forgotten except in the hearts of us, their lesser comrades, who understood their purposes and are made happier and more courageous by the memory of their examples.

These things are pleasing to remember, but discursiveness has been the ruination of this book up to the present time, and we feel that, at last, we should endeavor to get on with typhus.

Our discussions in preceding chapters have made it plain that there are no records of typhus fever in recognizable form in the ancient Oriental, Chinese, and classical literatures, and none in the chronicles and histories of the early Middle Ages. With the limitations of our own feeble learning, and with the good-natured assistance of a number of abler scholars, we have examined many of the accessible original records, and have studied the treatises of the leading medical historians. Fortunately for the amateur of epidemiological history, many profoundly learned men — foremost among them Schnurrer, Ozanam, Hecker, Hirsch, Murchison, Haeser, and Sticker — have gone over the ground with extraordinary thoroughness and have inserted into their works extensive citations of the critical passages from ancient writings. From them we have obtained, in addition to much information, guidance to sources, many of which were ac-

cessible in the Harvard Library, the Bibliothèque Nation-
ale, the Surgeon-General's library, and in the medical
libraries of New York and Boston. We cannot thus lay
claim to much originality in our literary investigations.
But we feel that there may have been some value in applying the criteria of contemporary knowledge to the scrutiny of ancient descriptions. None of the great historians
mentioned, though accurate and profound in their mastery
of languages and erudite in the medicine of their times,
had the assistance of the great mass of information
concerning infectious diseases which has accumulated
in the laboratories and clinics during the last thirty
years.

Applying modern technical judgment to the accounts
of infectious diseases of other times, we can find, in none
of the cases that have been cited as examples of typhus
fever before the twelfth century, trustworthy evidence
that the conditions described represented this disease as
it is known at the present day. The affliction of the
Clasomenian, the tenth and accurately described case in
the First Book of Hippocrates' *Epidemics*, cited as typhus
by Ozanam, appears to us more like a case of typhoid fever
than one of typhus. The only description in the *Epidemics*
which strongly suggests typhus is the one of Silenus,
which we have discussed at some length in a preceding
chapter. Not in Herodotus, Vegetius, Aëtius, or Galen,
nor in any of the other ancient writers who are cited, here
and there, as having seen typhus in classical and post-
classical periods, is there any description from which
reliable conclusions can be drawn. We might, from this,

with others who have had similar negative experience, deduce that the disease was actually new to Western Europe until shortly before the time of Fracastorius, that it was imported with soldiers from Cyprus, possibly preëxisting in a quiet way in the East. This, as we shall see, however, is not a necessary conclusion.

It will be helpful, before we go into this matter more deeply, to consider the descriptive criteria which justify us in assuming that any disease referred to by historians is actually typhus.

Typhus is an acute fever which does not always behave in a conventional manner. In its typical course it occurs more or less as follows: The onset may vary from extreme abruptness to a more gradual one. As a result the initial stages resemble closely those of severe influenza. The temperature rises rapidly, often to from 103° to 104° Fahrenheit, with chills, great depression, weakness, pains in the head and limbs. The eruption appears on the fourth or fifth day after the onset and, except in times of epidemic, the diagnosis is extremely difficult in the preëruptive stage. As the eruption appears, the fever is apt to rise. The rash usually begins on the shoulders and trunk, extending to the extremities, the backs of the hands and feet, and sometimes to the palms and soles. It becomes more abundant during the subsequent days, but it is seen very rarely on the face and forehead. It is at first composed of pink spots which disappear on pressure, but soon these become purplish, more deeply brownish red, and finally fade into a brown color. These are the "petechiæ" and "peticuli" of the older descriptions. A symptom of considerable importance, early and rarely missed, is the severe headache which is apt to be more unbearable in this disease than in other acute fevers; indeed, it is for this reason that one is inclined to assume, though not to assert with certainty, that varieties of the so-called *Kopfkrankheit* or *Hirnentzündung* of mediæval writers might have been typhus fever. Without the rash, however, and in the absence of an epidemic, the diagnosis of typhus fever would often remain uncertain even to-day, except for a specific reaction of the blood which was not available until quite recently.

When the rash, together with fever and headache, delirium and extreme weakness, is clearly described, typhus is easily recognized; but it must be remembered that the rash in the mild, isolated endemic cases — and especially among children — may be so slight and transient that often it is not noticed at all by the physician unfamiliar with the disease. For this reason, until typhus becomes epidemic, individual cases may often remain unrecognized, or may be described in such a general manner that it is impossible to differentiate them from measles, scarlet and typhoid fever, malaria, and a number of other febrile conditions that were common in ancient and mediæval times. Certainty that typhus existed in the fifteenth century and later is made possible largely by its epidemic occurrence. Under such circumstances, the description of individual, severe, and typical cases is reenforced by accounts of the characteristics of the epidemics, seasonal and other accessory factors, manner of spread, and mortality. Taken together, this information furnishes a structure of interrelated clues which permits certainty as to the nature of the disease.

We can thus conclude with some confidence that, as an epidemic disease, typhus did not exist in Europe until the fifteenth century. That Fracastorius and early Spanish observers regarded it as "new" will appear from the accounts of their observations with which we shall deal presently. It does not follow from this, however, that it did not occur at earlier periods as an endemic or sporadic, occasional fever — a smouldering source from which the later epidemic force evolved. That in its non-epidemic phase it should have escaped recognizable description would not be surprising. Among us, in the United States, typhus in this endemic form is constantly occurring. Yet until 1926, in spite of medical and educational resources far superior to those of earlier times, these cases remained entirely unrecognized. Have we, then, any basis other than pure surmise to assume that the disease is far more ancient than its epidemic history?

To answer this question, it becomes necessary to outline the natural history of the parasitism that is typhus, about which the last twenty years have taught us more than did all the centuries preceding. And this brings us at last to the consideration of the intimate family history, the immediate ancestry and the birth of the subject of this biography.

2

Until not very long ago, typhus fever was thought of as a single, individual disease, quite separable from other fevers, and unique. From studies, — none of them older than twenty years, and most of them carried on within the last six years, — we now know that typhus fever is the most distinguished member of a family of maladies which are grouped together, for reasons that will be clear presently, under the name of the Rickettsia diseases.

The kinships within the Rickettsia family may be outlined more or less as follows: In a position which we may compare to that of a stepbrother or maternal uncle stands Trench fever or Volhynian fever, which gave so much trouble to soldiers during the war and was conveyed to them by lice. The reason for placing this condition in a relatively distant relationship is the fact that in man it does not follow the clinical course which, in its basic manifestations, is common to all the other members of the family. We need not further pursue the fortunes of the Trench-fever branch, however, since it has little to do with the present discussion.

More closely related to typhus, quasi in the position of a second cousin, is Japanese River Valley or Tsutsu-gamushi fever. This disease is conveyed to man by the bite of the harvest mite, the *Trombicula akamushi*, and the insect picks up the infection from field mice and rats which are the natural reservoir of the disease. The virus is thus kept alive in endemic regions, by circulation between field mouse and harvest mite; and by the latter it is, on suitable occasions, transferred to man.

A closer relative, let us say a first cousin of typhus, is the disease — or the group of variants of the disease — called Rocky Mountain spotted fever. The infections

properly belonging to this division of the family are conveyed to man by the bites of ticks; and since, in these
cases, the virus can be hereditarily transmitted from both
the mother and the father tick to the little ticks, no
animal reservoir is necessary for continued survival. Yet,
since guinea pigs, rabbits, and a number of other animals
are susceptible to the disease, it is not impossible that
an animal reservoir, as yet undiscovered, may exist.

Probably identical with our American spotted fever is
the so-called, tick-transmitted, "typhus" of San Paulo,
Brazil. It is an interesting demonstration of the essential
similarity of these infections in man that the San Paulo
tick fever was regarded as true typhus by experienced
physicians as long as clinical observations unaided by laboratory study formed the sole criteria of judgment.

Another variant of the spotted-fever group is the
Fièvre Boutonneuse, or Escharo-nodulaire, which was
first described from Provence in the neighborhood of
Marseilles, but has also been found in Rumania. It is
tick-transmitted and, as in spotted fever, the virus passes
hereditarily from one generation of tick to another, without the necessary intervention of an animal reservoir.[1]

Finally, in true typhus fever we now know of two
distinct subfamilies, and suspect that others may exist.

As in the other Rickettsia diseases, the virus of both
varieties of typhus is transmitted to man by insects. The
body and the head louse carry the infection from one

---

[1] We omit, as having no direct bearing on the matter under discussion, any description of heartwater fever — a South African disease
of sheep, which is caused by Rickettsiæ and transmitted by ticks.

human being to another. The louse takes up the virus with infected blood, the Rickettsiæ multiply in the cells lining its stomach and intestinal walls, and appear in large numbers in the feces. Louse transmission was the great discovery made by Nicolle, which furnished the first powerful weapon for a counter-attack against the disease. It explained the manner in which epidemics are propagated. It removed all mystery from the historic association of typhus epidemics with wars, famines, and wretchedness. It justified the traditional designations of "camp fever," "prison fever," and "ship fever." But it left unanswered the problem of the persistence of the smouldering embers of the virus in interepidemic periods. For the human louse, probably a relatively recent host of the Rickettsiæ, is even more susceptible than man. It sickens and dies usually within twelve days, always within two weeks. Where does the virus persist between outbreaks? How are the interepidemic cases engendered?

An approach to the answer to these questions was furnished a few years ago by a study of the isolated cases of typhus which occur every year — here and there — in the United States. These cases occurred under conditions in which louse transmission could be excluded, and a search for other sources of infection was begun. The result was the discovery of typhus virus in rat fleas and then in the rats themselves. The epidemiological cycle seemed complete. Domestic rats carry the infection. In them it is perpetuated by transmission from rat to rat by rat fleas and by rat lice. Rat fleas will feed on man when driven to seek a new host by death of the old one — a frequent occurrence when domestic rats die or are killed. From the bite of the infected fleas the human being contracts typhus. This is the sporadic or endemic case. If the victim is lousy, group infection may result. If he lives in a louse-infected community, the consequence is an epidemic.

Since these facts were first ascertained in the Western Hemisphere, typhus-infected rats have been found in the Mediterranean basin, in places as widely separated as Syria, Piræus, Toulon, and North Africa; and so it is quite apparent that rat foci of the disease are widely distributed throughout the world.

But this is not yet the entire story. Mooser compared the strains of virus obtained from typhus cases occurring in European epidemic centres with those obtained in this country and in Mexico, and found that, although the two were as closely related as twins, they were not identical.

This differentiation has given rise to new problems and to the opinion, among some of us who are intimate with the family, that the classical European disease can maintain itself at all times in human beings and can persist without periodical rat passages. However, of this we shall have more to say presently.

To the lay reader, for whom this book is primarily intended, our catalogue of the Rickettsia family can hold little of much interest. Yet, without a survey of the family as a whole, it would be quite impossible to discuss the origin of typhus comprehensibly. The extraordinary aspect of the situation is the fact that, in one and the same era, mankind suffers from a group of almost indistinguishable acute fevers, which reach him by a variety of complex parasitic cycles, as follows: —

Tsutsugamushi	Mite →	Rat Field Mouse	→ Mite → Man
Spotted Fever Types	Tick →	Tick →	Man
Fièvre Boutonneuse	Tick →	Dog? →	Tick → Man

True Typhus

| Murine Type | Rat Flea
Rat Louse | → Rat →
(Mouse?) | Rat Flea → Man → Louse → Man |
| European (Human) Type | | Man → Louse → Man |

Were we engaged in writing a treatise for technical readers, this would be the place for emphasis upon the minor clinical differences between the members of the group — for such differences exist, as, for instance, in the necrotic local lesions of Tsutsugamushi, the glandular swellings in this disease, and the occasional raised knobs in Fièvre Boutonneuse. We might also enter upon details of the methods by which the individual strains of virus can be differentiated in the laboratory. But this would carry us, without much gain for our present purposes, too far from the central theme.

The fact remains that the family resemblances of these diseases in man are unmistakably close; are almost indistinguishably so in the spotted-fever–typhus relationship; and are demonstrable, as deep-seated biological kinships, by reactions of the blood of patients and by experimental observations upon infected animals. Moreover, all of the diseases of the group are caused by the invasion of the patient's body by the minute parasites spoken of as Rickettsiæ.

3

These minute, bacillus-like things belong to a group
which probably acquired its first parasitism on insects —
a surmise which is suggested by the frequency with which
similar organisms, incapable of causing disease in the
higher animals, occur in a variety of insects. Thus parasites
of this order have been seen in sheep lice, in dust lice, in
bedbugs, in mosquitoes, in fleas, in mites, and in ticks. The
name was given them by da Rocha Lima in honor of
Ricketts, an American who died of typhus while studying
the disease in Mexico City. The particular variety which
is responsible for typhus proper he called "Rickettsiæ
prowaceki," adding the name of Prowacek, an Austrian
who perished in the same manner. The Rickettsiæ needed
a name for themselves, because they cannot be logically
grouped either with the bacteria or with the Protozoa. In
the end, they will probably be found closely related to the
true bacteria. However this may be, for the present they
stand apart sufficiently to render a separate tentative classification convenient. They differ from true bacteria largely
in their response to ordinary methods of coloration, by
their refusal to grow on artificial media other than those
which contain living cells, and by the fact that in the living animal as well as in the tissue culture they multiply
only within the cell bodies themselves.

It is, of course, quite impossible to make even a reasonable guess regarding the free-living ancestral forms of
the Rickettsiæ. No doubt they were closely allied to true
bacteria. Indeed the characteristics by which the Rickettsiæ

differ from bacteria at present may well have developed as changes incidental to the evolution of their parasitic existences. At any rate, at some time in the very remote past, minute unicellular organisms became parasitic within a considerable variety of insects. In many cases they invaded the cells and became so adapted to intracellular existence that to-day they cannot be cultivated except in living tissue cultures.

We have few criteria by which we can appraise the antiquity of any form of parasitism. But in general, as Theobald Smith states it, pathological manifestations are only incidents in a developing parasitism. On this basis Rickettsia infection in the ticks is a very ancient condition; for in this relationship mutual tolerance has developed to such perfection that neither partner appears to be injured, and the parasite is transmitted, without harm to parent or offspring, from one tick generation to the next. In the rat flea the condition, though still ages old, is probably a more recent one; for the flea — after a month or two — gets rid of the parasite and recovers. In the case of the human louse, however, we are led by the same reasoning to assume a relatively late origin of the association. For no mutual tolerance has developed and the louse invariably perishes when infected.

The invasion of insects we may regard as the first step in that complex evolution which ultimately led to the human afflictions we are discussing. The next step was the transmission of the parasites from the insects to some of the higher animals. Some of the Rickettsia-infected insects belonged to species that had, themselves, become ectoparasitic upon animals and maintained themselves by sucking blood. In this manner Rickettsiæ gained access to those animals on which their insect hosts were in the habit of feeding. It is conceivable that the precise host channels through which the virus passed from insect to animal were dependent upon the accidental distribution of fauna in different parts of the world. Thus in one region it took the mite–field-mouse route, in another the flea-rat direction. And since in these two cases the mutual tolerance between parasites and hosts is still imperfect in both the insect and the animal phases, the virus is perpetuated only by an uninterrupted circulation of the parasite between the two. It is probably a fair guess that the tick-conveyed virus went through a similar animal-insect cycle centuries ago. It is even possible that a natural, but still unknown, animal host of spotted fever exists to-day. But the perfect adaptation which has made hereditary transmission possible within ticks has removed any necessity for an animal intermediary.

Thus we have a fairly reasonable basis for the tentative reconstruction of the natural history of the Rickettsia diseases. An insect-animal cycle once established, and given an insect which, in emergency, will feed on human subjects, the transfer of the parasites to man follows.

Man is, in the biological sense, a recent host, and in him Rickettsia invasion arouses a physiological resentment. A struggle between invader and host ensues which manifests itself as disease. One or the other succumbs. But for the parasite it is a Pyrrhic victory. When the man dies, the Rickettsiæ that have killed him die with him. Only those survive which can escape into a louse, or possibly a flea, which happens unwisely to feed upon the human victim at a time when the Rickettsiæ are circulating in the blood. And, of the two, the louse is by far the more dangerous — in relation to epidemic spread; for although, unlike the flea, it can neither hop nor live for any length of time separated from its human host, it possesses qualities of dogged persistence and patient diligence which arouse that admiration, thinly masked by a pretense of loathing, which men similarly feel for competing races whom they fear and, therefore, persecute.[2]

To those who are engaged in the technical study of the typhus group it is apparent that the facts so far ascertained concerning the insect-animal parasitism of the Rickettsiæ represent only a beginning. Apart from the practical importance of these relations in their bearing on diseases, they offer to the general biologist a rich field for the study of parasitic cycles. It is quite likely that Rickettsia invasions have taken many directions other than those so far investigated. In the Malay States, Formosa, Sumatra, and Annam, perhaps also in Japan, the Tsutsugamushi virus can pass through rats as well as mice; and in the same places, together with a flea-born true typhus, there is also a tick disease. These are being unraveled by investigators all over the globe. It has been shown experimentally that by artificial inoculation virulent Rickettsiæ can be kept alive for a week or two in a number of insects that do not naturally harbor them. Also many species of

[2] We refer to the "Blond Aryan" complex.

animals — not as yet convicted of being sources of human disease, such as domestic mice in Europe and America, New World mice, rabbits, woodchucks, monkeys, and even horses and donkeys — can be inoculated with the Rickettsiæ and harbor them for varying periods. In many of them this maintenance of the virus is peculiarly dangerous because it is what we call "inapparent" — that is, the animal shows no signs of illness, yet retains within its body a virus capable of transfer to insects or to other susceptible animals. "Inapparent" infection is beginning to possess an importance of the first order in epidemiological reasoning in many fields other than that of typhus fever. In the Rickettsia problems, however, it has already attained practical significance. A rat inoculated with typhus virus shows no apparent symptoms except, in some cases, a little fever. Yet two or three weeks later one can produce typical typhus reactions in guinea pigs or infect lice by intrarectal inoculation of the apparently healthy rat's brain! But this is again tempting us into discursiveness. We return to our main theme.

CHAPTER XIII

In which we consider the birth, childhood, and adolescence of typhus

I

THERE are, as we have stated, two distinct types of true typhus virus. The diseases they cause in man are identical and both are transmitted from one individual to another by human body and head lice. Both in man and in animals recovery from one type protects against the other, testimony of their close and fundamental kinship. They can be distinguished only by relatively slight but definite differences of behavior when inoculated into guinea pigs, rats, and mice, and by reactions, called immunological, which are far too technical to occupy us here. Before these distinctions had been recognized typhus had been regarded all over the world as a single disease perpetuated by man-louse-man transfer. This observation, however, together with epidemiological observations in Australia and American case studies, led to an intensive search for virus reservoirs other than man. The result was the discovery of natural rat infection and of rat-flea transmission.

Now in correlating the origin of virus strains with their manner of behavior in guinea pigs, it was soon observed that all the viruses obtained either directly from rats or from rat fleas, as well as those isolated from human victims in America and Mexico (regions where the presence of infected rats and epidemiological circumstances indicated rat origin), behaved in one way; while the strains obtained from man in Southeastern and Eastern Europe — where endemic and epidemic typhus has been prevalent for centuries — behaved in another manner. For these reasons students of the disease to-day classify the two varieties as the "murine" type — in which the rat-flea cycle precedes human infection — and the classical or "human" type, for which no rat origin has as yet been determined. The precise relationship between these closely allied subvarieties then became the focus of attention, since it was obvious that comprehension of this would go far toward explaining the epidemiology of the classical European disease — thereby furnishing new principles for protective measures. The speed with which things have been moving in the typhus world may be gathered from the fact that most of the work we are discussing has been done since 1928, a good deal is hardly off the presses, and some of it is not yet in print as these paragraphs are being written. In its accomplishment, French, Swiss, American, British, German, Mexican, and Polish investigators have engaged in the sort of exciting, friendly, and eager competitive collaboration or collaborative competition which gives our profession a zest and charm and a freedom from nationalistic chicane found in few others.

It was necessary, first of all, to determine whether the two types were permanently fixed in their differential characteristics or whether they represented temporary variants — or, as they are now called, "dissociations" of one and the same virus, dependent upon or induced by the different hosts through which they passed. This question has, in our opinion, been answered — though in the interests of an accuracy that is essential even in a superficial discussion of such things we must add that there is still an element of speculation in the explanation, and opinions are not yet entirely unanimous. In approaching the problem, investigators began to pass both types of virus through a variety of insects, through guinea pigs, rats, and mice, and to collect for study all the strains they could get hold of from rats and from human patients. As the matter stands, at the end of about five years of such study the evidence so far accumulated tends to show that the two varieties are permanently fixed — each in its own form. They have so many overlapping characteristics — even in the animal experiments — that it is quite easily possible to train one of them into a temporary simulation of the other by special methods of investigation. But as soon as there is a relaxation of experimental manipulation each type snaps back into its original condition. There are strains of the murine and of the European type in American and foreign laboratories that have been so observed for three, four, and five years and are still true to type.

We can assume with much confidence, therefore, that the two varieties are fixed, though very closely related, variants. But the ease with which one of them can temporarily be trained in the direction of the other by experimental manipulation suggests that the differentiation is one that has come about, biologically speaking, at a relatively recent period. Some light on this phase of the matter has come from accidental observations made on strains obtained in Mexico. Now and then, from one and the same Mexican epidemic, among typical murine strains, a few aberrant ones have been recovered which act like the European or human variety. Some of these may retain their human strain characteristics through many guinea-pig passages. Eventually, however, all of them, especially under the influence of rat passage, have "come back" to the murine attributes. Since in the Mexican epidemics the passage from man to man — just as in the Continental epidemics — is a louse transmission, the observation just cited suggests that passage through man and lice tends to modify the properties of the murine virus into a closer similarity to those of the European human type.

Since years of animal passage and selective experiment have failed to produce a reversion of a human virus toward the murine, — whereas passage of a murine through man rapidly produces an often obstinate, though so far still temporary, change in the direction of the human, — we have much reason to suspect that the human is an off-shoot of the former — the murine being the original typhus virus of man, which, after a sufficient number of man-louse-man passages, becomes stabilized as a slightly changed but permanent and fixed variety. Under such circumstances, we may still ask ourselves whether the classical European virus is renewed, from time to time, from rat sources and so perpetuated; or whether, on the other hand, it has become thoroughly and permanently established in man and is continued between epidemics by a trickle of man-louse-man cases or by so-called human carriers who maintain the virus for long periods, though appearing completely cured — much as in the case of the "inapparent" animal infections which we have elsewhere described.

A partial answer — in our view a complete one — to this query has come from the study of European cases imported to America. There occurs among the crowded immigrant population of our Northeastern cities an acute fever called "Brill's disease," which is really typhus and yields a typical European or human virus. When Brill first described it in 1898 among Jews in New York, being unfamiliar with typhus, he thought it a "new disease." We mention this in no disparagement of an extraordinarily sagacious physician; but rather because, if errors of this kind are easily made in the present era of medicine, we must be doubly careful in appraising remote historical evidence bearing on the antiquity of infectious diseases. Brill deserves much credit for having differentiated these mild cases from similar fevers then prevailing, and calling attention to them. His error, moreover, has been a common one in the history of medicine. As Murchison tells us, "So completely did relapsing fever disappear from Britain after 1828 that when, after an interval of fourteen years, it again showed itself as an epidemic in 1843, the junior members of the profession failed to recognize it and it was regarded as a new disease." Many similar instances could be cited.

But to return to Brill's disease. This, as we have said, is European typhus brought to this country by immigrants from the typhus regions of Southeastern Europe. It is not common, but there have been enough cases to permit profitable study. More than five hundred of them are on record as occurring in Boston and New York since 1910. Epidemiological analysis has shown that well over 90 per cent of all of these cases occurred in the foreign-born, although they lived in close association with their native-born friends and relatives, and had similar customs. The cases were so distributed in time and place that louse transmission or contact infection could be excluded, and the circumstances of over five hundred carefully investigated patients showed that no factor common to the entire population — such as rats or fleas or any other animal or insect vector — could be held responsible. To make a long story short, the investigations showed that these cases were, almost all of them, recrudescences of infections acquired in childhood in the native heaths of classical typhus, and that the classical European typhus can maintain itself in human reservoirs indefinitely without the intervention of extraneous animal vectors.[1]

The situation, in summary, is the following: There are two very closely related, but nevertheless distinct types of typhus fever prevalent side by side on both the American and the European continent. From suggested, but yet incomplete information, one is inclined to assume that probably the two types exist in many other parts of the world. One of these varieties, which we speak of as the murine virus, is maintained in interepidemic periods in

---

[1] The discussion of the prolonged survival of an infectious agent in the bodies of convalescent and recovered men and animals would carry us into a new, long, and complex chapter. And we have set our self-control firmly against further digressions.

rats, possibly in mice, where it passes from one animal to the other by the insects we have named; occasionally it gets into man with the bite of a rat flea; but it causes group infection or epidemics only when the circumstances are such that human lice can transmit it from man to man. The other type has become solidly established in man. Some individuals who have recovered from a first attack retain the virus in their bodies and may have another attack of the disease many years after the first one, when their resistance is depressed for reason that it has not yet been possible to analyze. From these recrudescent cases epidemics can start under conditions of general louse infestation. There are quite probably rat and human reservoirs side by side in many different parts of the world, but a complete survey of this situation will probably take a good many years of further study.

2

We have now reached the point in our biography when we can speak of the birth of our hero without fear of being forced into further explanatory digressions. If hitherto we have followed the discursive plan of Dr. Sterne in *Tristram Shandy*, we can insist — and the reader will agree with us — that we were not impelled, as was the immortal author of that great work, by a desire to be humorous, but rather by the nature of our subject. The birth of an infectious disease is not as simple a matter as that of a man. Gestation is not a mere matter of ten months or so, but represents complex biological inter-adaptations and interactions which cover thousands of years. And in this particular case we may say that the *conception* of our disease took place when the first Rickettsiæ became parasitic on insects; and *gestation* lasted through the uncertain but undoubtedly centuries-long period during which the parasitism progressed from insect to animal, and finally through other insects to man himself.

Under the circumstances described it appears probable that isolated, endemic, rat-to-man or mouse-to-man cases of typhus occurred centuries before the disease became epidemic, recognized, and differentiated. It is almost sure that wild rats and possibly other rodents were infected in many parts of the world early in the natural history of this parasitism. In the Malay States to-day there seems to be a concentration of endemic cases of tropical typhus among workers on oil-palm plantations, where rats abound.

Although, in attempting to postulate a preëpidemic existence of typhus fever before the fifteenth century we are fishing, to some extent, in speculative waters, there is still much to support this view in conditions as they exist in widely separated areas of the world to-day. In Mexico and in the Southern United States a dribble of sporadic cases, long unrecognized, are constantly acquired from domestic rats, which result, in the former country, in epidemic outbreaks only when the louse takes a hand. In Malaya — where there is an urban and a rural tropical typhus — the diseases are again sporadic and rarely give rise to group infection. The rural variety which attacks the workers on the oil-palm estates seems to be a danger chiefly for those laborers who are occupied in the clearing of the lalang growth and the weeds at the bases of the trees. They are exposed to some vector — possibly wild rats and their fleas, or some still undiscovered vector which lurks in the brush. At any rate the virus is here widely distributed in nature, evolved completely up to the point at which it is ready to enter man — and has probably so existed for an indeterminably long period. Similar conditions prevail in regard to Tsutsugamushi. Incidentally, failure of epidemic outbreak of the typhus in Malaya is probably due to the fact that, according to Dr. Enid Robertson, the body louse is almost unseen in Malaya, though head lice exist. In hot countries, where men are totally unclad, or are clad lightly, and groups of people are widely scattered in rural settlements, the chances are great that typhus will remain endemic almost permanently and become epidemic only when conditions of living are modified.

In problems concerning the remote origins of diseases there is little chance of either proving or disproving any hypothesis. We believe, however, that the biological observations to which we have devoted much space strongly suggest the following tentative theory regarding the pre-epidemic history of typhus.

Typhus fever was born when the first infected rat flea fed upon a man. This accident probably took place — most likely somewhere in the East — centuries before the disease reached the crowded centres and the armies of mediæval Europe. Endemic and usually mild cases occurring here and there, with rarely a group outbreak, escaped the attention of ancient physicians and historians — or were not differentiated from other febrile diseases.

The murine virus was thus the original *typhus*. In the course of time the disease was carried, perhaps repeatedly, to Western countries — chiefly by armies, at first causing limited outbreaks that perhaps ended with the virus still remaining largely or entirely murine. Infected rats became established in the Mediterranean basin. Early localized epidemics thus may — like those in Mexico to-day — have remained murine in origin for a long time. And indeed, in these earlier days of its epidemic history, typhus outbreaks were relatively far apart. In the sixteenth and seventeenth centuries, beginning with the campaigns of Maximilian against the Turks and through the Thirty Years' War, the disease became an almost incessant scourge of armies and was scattered far and wide among the wretched populations under conditions — ideal for typhus — of famine, abject poverty, homeless wandering, and constant warfare. The human louse was possibly the last of the series of hosts to acquire the virus — for it had, long before this time, become inseparably dependent upon man. And this surmise is in keeping with the fact that in the louse the Rickettsiæ are more predatory than parasitic. The infected louse always dies.

Under the conditions which we have described for the unfortunate centuries it is quite conceivable that typhus fever may have been almost uninterruptedly propagated by the man-louse-man route in certain parts of Europe, with renewal at any time from a rat-flea source (although endemic rat-transmitted cases may have been occurring at the same time). And this continued through the eighteenth century, which is, par excellence, the Century of Typhus.

Thus constantly passing through lice and men, certain strains became modified — even as, in a less permanent manner, we can observe such changes after a few man-louse passages in modern Mexican outbreaks. Thus was born the younger brother — the human virus. The two persist, side by side, in many countries of Europe, and, as investigations of Brill's disease have shown, here among us in America as well — the murine brother having its permanent home in rats and fleas; the human, last born, firmly established in man.

The preëpidemic history of our disease, the circumstances of its birth and adolescence, are and needs must remain largely hypothetical. We have constructed a trellis of likelihood from known facts concerning the natural history of the virus. What we may call the adult state of the disease — the period at which it became a powerful factor in the history of mankind — began when it acquired epidemic propensities. Then only was it recognized as an individual and accurately described, and we are again on the *terra firma* of reliable information in our next chapter, which deals with the vigorous young adult phases of our hero.

CHAPTER XIV

*In which we follow the earliest epidemic exploits of
our disease*

I

WE assume then that the original Rickettsia parasitism
which led to typhus fever in man was a rat-rat-flea infection, and that this gradually infiltrated into Western
Europe from the East. This parasitism exists to-day in
and around the Mediterranean basin, widely distributed,
and there is no particular reason to believe that it got
there from American foci. At first the disease in man
probably bore the form of the mild, sporadic cases —
scattered in time and space, as they occur in the Southeastern United States to-day.

Considering the state of medicine in the early Middle
Ages (possibly cases appeared as early as the Crusades),
we cannot expect records of any value. For, as we have
seen, the existence of the disease among us was not
recognized until quite recently, and the diagnosis, even
now, calls for considerable skill and experience in these
relatively benign infections, in which the fever is often
short-lived and the rash so insignificant that it may be
overlooked entirely or mistaken for flea bites.

Early group infections, when they occurred at all,
probably did not extend beyond the limited ranges of
family or village association. And when, in its earliest

epidemic appearances, the disease attacked armies or towns, there is much reason to believe that it was associated with a number of coincident infections — plague, enteric fevers, scarlet fever, measles, and so forth, and was obscured, in historical records, in the general undifferentiated mess of "pestilence." The conditions which let down the bars for one type of infection usually admit a great many others; and, except under special conditions, epidemics are usually composed of a number of different types of transmissible disease.

In the East it is probable that typhus had passed from the endemic to the epidemic state at a period earlier than it did in Europe, and there is some reason to assume that the earliest recorded severe European epidemic was transported with soldiers from Cyprus to Spain. This epidemic occurred in 1489 and 1490, when the forces of Ferdinand and Isabella were at grips with the Moors for the possession of Granada.

Of considerable significance for our view of the gradual manner in which typhus became epidemic in Europe is the fact that we have information that can hardly be questioned of at least one preceding group infection which occurred some four hundred years earlier in a monastery near Salerno. It is described in the *Cronica Cavense*, which we have not been able to see in the original, but from which Renzi, much quoted by medical historians, has cited the important passages. Through the kindness of Major Hume of the Army Medical Library, we have been able to find the following passage taken from the *Storia di Medicina in Italia*, Volume 2, Napoli, 1845). "E fra'

tanti esempi ne prescegliero uno abbastanza antico per
potere dissipare ogui dubbiezza. Nella Cronica Cavense
reportata dal Pratillo (tom. 14, pag. 450) leggesi: *Anno
1083 in Monasterio Cavensi in mense augusto, et septem-
bri crassavit pessima febris cum Piticulis et parotibus.*
Nel che si ravissa chiara la differenza che si metteva fra
la pesti, la febbre di altro genera, e quella accompagnata
da petecchie." ("In the year 1083, in the monastery of
La Cava in the month of August and September, there
spread a severe fever with peticuli and parotid swellings,
in which one sees clearly the difference which is found
from the Pest, a fever of a different kind and — in this
case — accompanied by petechial spots.") From this passage it seems that a diagnosis is warranted.

It would be strange if there had been no typhus whatever between this outbreak and that of 1489. We are almost compelled to assume that, during the interval, no
accurate observations were recorded, or that, at any rate,
if made, they have been lost.

The chief source for information of the early epidemics
of Spain is the book by Joaquin Villalba, which bears the
following title: *Epedimiologia española o historia crono-
logica de las pestes, contagios, epidemias y epizootias que
han acaecido en españa desde la venida de los cartagi-
neses, hasta el ano 1801. Con notioia de algunas otras
enfermedades de esta especie, etc. Madrid, en la imprenta
de Don Mateo Repulles, 1802.*

Villalba derived much of his information from a work
in the title of which the word "tabardillo" is first applied
to the disease. It is *De febris epidemicos, et novos quos*

latine punticularis, vulgo tabardillo et pintas dicitur, natura, conditione et medela. It was attributed, by Nicolas Antonio and Alberto de Haller, to a certain Alonso de Torres. Villalba believed that the true author was Luis de Toro, who wrote at the instigation of the Marques Don Luis de Astuniga y Avila. Avila, realizing that the history of this disease had never been written, wanted it recorded. The first reference to a typhus epidemic which occurs in Villalba's book is the following: —

Among the important epidemics which are referred to by our historians, there is one which began during the civil wars of Granada, in the years 1489 and 1490. Later, this disease spread among the Spaniards, as we shall see in the discussion of the plague of 1557. This disease was a malignant spotted fever believed by some to have originated from the unburied corpses; by others assumed to have been introduced by soldiers who came to the Granada wars from the island of Cyprus — an island in which this fever was prevalent. In Cyprus, these soldiers fought with the Venetians against the Turks, and thence they carried the seeds of the disease not only to the Spaniards, but also to the Saracens. However this may be, the physicians of that time believed that the spotted fever was contagious and identical with plague.

The disease of which we are speaking was disseminated from the camps of Granada to the army of Don Fernando the Catholic. Whether for this or some other cause, when the army was reviewed at the beginning of the year 1490, the generals noticed that 20,000 men were missing from the rolls, and of these 3000 had been killed by the Moors and 17,000 had died of disease, not a few of them succumbing to the severe cold — a kind of death which, says Mariana, was very miserable.[1]

[1] *Entro las epidemias notables que se refieren por nuestros historiadores, es la que tuvo principio en tiempo de las guerras civiles de Granada,*

There can be little question that this was typhus fever, and one of the most interesting parts of the passage is that in which the origin of the infection is referred to "certain soldiers who came to the war of Granada from the Island of Cyprus, to which island this disease is peculiar. . . ."

In the second paragraph the disease is credited with having killed 17,000 soldiers, as against the 3000 killed at the hands of the Moors.

In the next passage, which deals with the epidemic of 1557, Villalba again indicates that the disease was newly imported at the time of the civil wars for Granada. By this time, epidemics had spread over the entire Spanish peninsula and raged, unchecked, for thirteen years, until 1570: —

A new disease, unknown until the time of the civil wars in Granada, appeared in Spain in the year 1557 and depopulated the greater part of our peninsula; it did not begin to decline until

---

acaecidas por los anos de 1489 y 1490, cuya enfermedad se comunico despues a los españoles, como veremos al tratar de la peste de 1557. Esta enfermedad fue una calentura maligna punticular, nacida de los cadaveres insepultos, segun algunos; o traida, segun otros, por ciertos soldados que vinieron de la isla de Chipre a la guerra de Granada, de cuya isla era peculier esta fiebre, donde pelearon contra los turcos a favor de los venecianos, y conduxeron el seminio de este mal contaminando no solo los españoles sino tambien los sarracenos. Como quiera que sea, juzgaron los medicos de aquel tiempo que la fiebre punticular era contagiosa y nada agena de la naturaleza de peste.

Ya sea que la peste de que acabamos de hablar se comunicase de los campos de Granada al exercito de Don Fernando el Catolico, o bien por qualquiera otra causa, al pasar revista de el a la entrada del ano 1490 hallaron los xefes militares, que faltaban en las listas veinte mil hombres, los tres mil muertos a manos de los moros, y los diez y siete mil de enfermedad, y no pocos por la aspereza del invierno se helaron de puro frio, genero de muerte, dice Mariana, muy desgraciado.

the year 1570. This new pestilence was believed to have originated among the Saracens after the war of Granada; that is,
after the King Don Fernando of Aragon and Doña Isabel,
Queen of Castille, conquered that city; and after the Moors had
been dispersed by the decree of Don Felipe II. That this infection came from the Spanish Arabs can be deduced from the fact
that almost all those who were driven from their homes infected,
by association and contact, the inhabitants of the villages, towns,
and cities, as related by Luis de Toro in his treatise *On Spotted
Fever;* in this work may be found his description of the character
of the disease as it occurred in the periods 1570 and 1577.[2]

In his description of this particular outbreak which is
too long to be quoted in full, Villalba expresses the belief
that the typhus of America originated at this time by transportation from Spain to Mexico. The passage in question
is the following: —

This spotted fever — which, as we have said, afflicted the people
of Spain — was transferred to America with our navy and our
merchantmen and attacked the noble city of Mexico with such
severity that it caused much distress. Dr. Francisco Bravo, native
of Osuna, and physician of that city, wrote an extensive discourse

[2] *Una nueva enfermedad desconocida de los siglos antiguos hasta las
guerras civiles de Granada aparecio en España el ano 1557, la qual
despoblo la mayor parte de nuestra peninsula, y no empezo a corregirse
ni mitigarse sino hacia el ano 1570. Esta nueva petilencia se cree que
tomo su origen de los sarracenos despues de la guerra de Granada; esto
es, despues que el rey Don Fernando de Aragon y Doña Isabel, reyna
de Castilla, conquistaron dicha ciudad, y despues de haber sido dispersados los moriscos por decreto del señor Don Felipe II. Que esta in-
feccion proviniese de los arabes españoles, se colige de que casi todos
los que fueron dispersados, inficionaban con su comunicacion y trato a
los habitantes de las aldeas, villas y ciudades, como refiere Luis de Toro
en su tratado "de febri punticulari"; a suyo caracter pertenece, y se
hallara su descripcion en la epoca 1570 y 1577.*

about this disease, to which he gave the name of "tabardete." His rare work is entitled: *Opera Medicinalia in quibus quam purima extant scitu medico necessario, in quatuor libros digesta,* printed in Mexico by Pedro Ocharte in the year 1570, in octavo. This work, dedicated to Prince Don Martin Enriques, contained the description of this disease, causes, signs, and symptoms and cure — together with other considerations to which we shall refer at the proper time.[3]

On this subject we shall have more to say presently. The disease was thus well launched in an epidemic form in Europe during the last decade of the fifteenth century and throughout the sixteenth, but had not yet spread widely across the Continent. In 1546 Fracastorius published his *De Contagione,* in Chapter VI of the Second Book of which he gives an excellent description of the disease in its clinical appearances and has many sagacious things to say about its nature and the manner of its spread. The following passage from the beginning of this chapter is cited after the translation of W. C. Wright: —

There are also other fevers, which, in a manner of speaking, come midway between the truly pestilent and the non-pestilent, for though many die of them, many recover. They are con-

---

[3] *Esta fiebre punticular, que desolaba los pueblos de España, como acabamos de decir, paso a las Americas por medio de nuestras naves y comercio, y acometio con tanto rigor a la insigne ciudad de Mexico, que causo en ella no pocos estragos. El doctor Francisco Bravo, natural de Osuna, y medico de aquella ciudad, escribio sobre ella un largo dis-curso con el nombre de tabardete, que se halla en su rarisima obra titulada: "Opera medicinalia in quibus quam purima extant scitu medico necessarie, in quator libros digesta," impresa en Mexico por Pedro Ocharte ano 1570, en octavo. Esta obra dedicada al principe Don Martin Enriquez, contiene la descripcion de esta mal, causas, senales, sintomas y curacion de el, con otros tratados, de que caremos noticia a su tiempo.*

tagious, and hence partake of the nature of pestilent fevers, but they are regularly called malignant rather than pestilent. Of this sort were those fevers which in 1505 and 1528 appeared for the first time in Italy, and had not been previously known there in our time. They are, however, familiar in certain parts of the world, for instance in Cyprus and the neighboring islands, and were also known to our ancestors. They are vulgarly called "lenticulæ" (small lentils), or "puncticulæ" (small pricks), because they produce spots which look like lentils or flea bites. Others spell the name differently, and call them "peticulæ." We must study them carefully, because nowadays, too, they are frequently observed, not only as affecting many at once, but also as special cases, in individuals. Instances have been observed of persons who went from Italy to other countries where no fever of this sort existed, and died of it there, as though they had carried the infection with them. This happened to that very celebrated and learned man Andrea Navagero, ambassador from the illustrious Republic of Venice to the King of France, some years ago. For he died of this disease in a province where that sort of malady was not known, even by name. He was a man of such learning and genius that no greater loss to letters has been incurred for many a year.

We are quoting copiously from the writers who observed these early epidemic appearances of typhus in Europe, because we wish to emphasize the fact that it was regarded, in this form, as a new disease and there was general agreement that it came to Europe from the East. Of course, in this respect, they may well have taken their view from the early opinions expressed by Luis de Toro, and his opinion concerning the transportation of the disease from Cyprus may have been wrong. Infected rats are now present along the southern border of the Mediterranean. Spain may have been the first Continental region to be attacked because of active communication across Gibraltar and the rapid spread of rats from one continent to the other.

However this may be, before the middle of the sixteenth century typhus fever had begun to take an active hand in the politics of Europe. It made its political début, as one may call it, by one of the most far-reaching and profoundly effective strokes of its entire career, playing the decisive rôle in the relief of the Imperial army at Naples when besieged by the French under Lautrec in 1528.

We may best appraise the enormous historical importance of the short and localized epidemic of typhus which destroyed the French army before Naples by considering the background of the political conditions.[4] Northern Italy was the battleground on which Charles V and Francis I had long disputed the hegemony of Europe. The key to the situation was alliance with and power over the Pope. On February 24, 1525, the victorious march of the French army was turned into utter defeat when the Spanish troops and their German allies, led by Pescara, snatched victory from imminent defeat. Italy fell to the mercy of the Imperial army and the French King became a prisoner in Spain. The Pope, Clement VII, was in a difficult position. He feared for the independence of the Holy See: with Milan and Naples in the hands of the Emperor, the papacy was completely encircled. Lannoy, the most energetic of the Imperial generals, was threaten-

[4] See Von Pastor, *History of the Popes.*

ing to march on Rome. The Pope was forced to advance
huge sums of money and to enter into an alliance with
the Emperor. In 1526, after the Peace of Madrid, Francis was liberated. The conditions imposed upon the French
King were so severe that historians find it difficult to
understand how so astute a monarch as Charles V could
ever have expected him to remain true to his undertakings. Again, the Pope — who was timid by nature —
fluctuated between two terrors: one, the immediate fear
of the Imperial power in Italy; the other, apprehension
of the consequences if, having remained true to his alliance with Charles V, he should soon be confronted with
a returning French army. Added to his European difficulties, the rapidly advancing power of the Turks in the
East, with invasion of Italy threatened by way of Apulia,
conspired to confuse the papal diplomacy. In 1522 Rhodes
had fallen to the Moslem power. One of the chief bul-
warks on the Eastern front was thus destroyed, the Turks
were in Belgrade, and in 1526 they had destroyed the
Hungarian army at Mohács.

Although he desired to establish armistice and to remain neutral, the distracted Pope was nevertheless per-
suaded to throw in his lot with Francis I, and the result
was the League of Cognac, formed in May 1526, between
Clement VII, Francis I, Sforza for Milan, and the Republic of Venice. Active warfare — which, as a matter
of fact, had never entirely ceased — flamed almost immediately. Francis I, enjoying his new liberty, was slow
in sending assistance, and the Duke of Urbino, who
commanded the northern armies of the new League, was

excessively timid in his tactics. The consequence was that Milan and Siena remained in Imperial hands; and while the Pope was sending urgent messages to Francis I for assistance, the Colonna fell upon Rome, and with a small army of 5000 men drove the Pope into the Castello Sant' Angelo, sacked the city, — including the Vatican, whence they took the papal tiara, — broke into the secret chapels of St. Peter's, and — before retiring — inflicted a damage which is estimated to have amounted to 300,000 ducats. The Imperial armies under Frundsberg and the Duke of Bourbon, soon after this, marched southward in Italy and approached Rome. The first attacks upon the city were made in May 1527.

There followed the sack of Rome — one of the most dreadful calamities that, in its long history, had befallen the Sacred City. The Pope was made a prisoner. The conditions in the city were described by a Spaniard, Villa, as follows: "In Rome no bell sounds; no church is open; no Mass is read. There are no Sundays and no holidays. The rich shops of the merchants are used as stables; the most beautiful palaces are devastated. Houses burn, and the streets are heaps of manure. The stench of the corpses is dreadful, and in the churches I have seen dead bodies gnawed by dogs. Mercenaries are dicing for heaps of ducats in the streets. I can compare it to nothing that I know except the destruction of Jerusalem." The captivity of the Pope was dreadful, not only for physical suffering and anxiety, but also because it was aggravated by an outbreak of plague which came with the summer and killed enormous numbers of the citizens, including many of those immediately attached to the Pope's person. Two cardinals, imprisoned with him, died of the disease, which was probably bubonic plague.

The same disease — contracted in Rome — killed Lannoy, the Imperial general. The death of this energetic leader probably had a good deal to do with the subsequent initial successes gained by Lautrec, who led the French troops that were now approaching Northern Italy. At first, the advance of the French was a triumphal march. Lautrec, to whose French corps were added mercenaries from Lorraine and the Rhine, and constant reënforcements of Italians, who regarded him as a liberator, reconquered the cities of Lombardy almost without opposition, and learned of the liberation of the Pope and his transfer to Orvieto when he reached Bologna. Meanwhile, the Spanish troops, delayed by the pleasures of sacking Rome, at last became alarmed. Realizing that a decisive battle would have to be fought for Naples, they hastened to fortify the place, largely through the advice of the Prince of Orange, who foresaw and notified Charles V of the perilous situation.

The Imperial army, which had taken Rome, had by this time been reduced, largely by the plague, to less than 11,000 men, and was wild and undisciplined. This débris of the once powerful army was encircled near Naples, at Troja, by Lautrec, with about 28,000 men. Unfortunately, Lautrec did not immediately attack, but gave the Prince of Orange an opportunity to escape during the night and fortify the position at Naples. It must be remembered that, at the time Lautrec's army arrived before Naples, war upon Charles V was being carried on in all the territories of his possessions — in the low countries, in Catalonia, and along the Mediterranean coasts. On April 28, the Imperial navy was almost destroyed, and by June 10 Genoese galleys completely blocked the harbor of Naples. On the fourteenth of June, 1528, after a month and a half of the siege, the Prince of Orange wrote to Charles V: "For ten days we have been living on bread and water; meat and wine have failed us, and your troops have not been paid for a long time." He added: "Neither they nor I can accomplish the impossible, and when another month has passed, we will be near the end."

It is impossible to estimate the consequences for the future history of Europe if Naples had fallen at this time, with Italy and the Pope ready to acknowledge Francis I as liberator and defender of the faith — but then came Typhus. On July 5, Lautrec had believed Naples incapable of resisting any longer, but in the marshy, crowded camps of the French the pestilence was destructive and rapid. Within thirty days, more than half the army died; according to some accounts, of 25,000 men only 4000 remained. Vaudemont, Navarro, and Lautrec himself were taken sick and died. Their successor, the Marquis of Saluzzo, realized that the siege must be immediately raised. On a rainy night of the twenty-ninth of August, the retreat began, closely followed by the energetic Prince of Orange with his cavalry. The remnants of the French army were cut to pieces. They were murdered or disarmed, to perish later at the hands of the peasants. A few bands managed to reach Rome, half naked and sick. The Emperor was completely triumphant, and Clement VII made overtures. The dependence of Italy upon Spain and the complete control of the immense influence of the papal power by Charles V was fully established. In 1530 Charles V was crowned ruler of the Roman Empire at Bologna, by the power of Typhus Fever.

2

Villalba, in a passage which has been cited, suggests that typhus was transported from Spain to the New World during the first half of the sixteenth century.

Ever since the discovery of the New World by the Old there has been an interchange of many things, good and bad, between them. At first it was a very uneven exchange. The Old World brought culture and small-pox, the Christian religion and measles, rum, European quarrels, scarlet fever, sparrows, horses and donkeys, Anglo-Saxons, Irishmen, Jews, Negroes, trousers, influenza, wheat, brotherly love, gunpowder, and tuberculosis. For all these blessings it received in return at first only gold, tobacco, syphilis, potatoes, and Indian corn. As the New World flourished it began to pay a more adequate interest on the invested capital. At present the honors are about even. Some of the things America has received from her elders, like industry, politics, capitalism, Communism, alcoholism, methodism, baptism, free verse, free love, psychoanalysis, educational systems, journalism, philanthropism, the camera, science, art, literature, football, rats, remittance men, gypsy moths, Russian princes, starlings, macaroni, Wiener Schnitzel, labor troubles, bankers and brokers, and so forth, and so forth, we repay either in kind or in a bigger and better way. And we add for good measure high tariffs, peanuts, phonographs, chewing gum, moving pictures, breakfast foods, heiresses, Christian Science, cocktail shakers, efficiency methods, and the boloney dollar. Yet in many ways we shall always be a colony of Europe, since in the cultural storehouse of two thousand years there are gifts for which we have no coin to pay. But this is again aside from our subject. We are interested in the present connection in whether typhus fever existed on the Western Hemisphere before these regions were discovered by Europe, or whether this too is an importation.

The disease in a form somewhat different from that of Europe and of Africa is at present prevalent in Mexico, Peru, Brazil, Bolivia, Chile, and in the Southeastern and Middle Eastern United States. Its close relation, Rocky Mountain spotted fever, is at large — as we shall see — in the central plateau and mountain regions of our republic, and probably in many of the other countries mentioned. In Mexico typhus has existed for several centuries. Was it brought by the *conquistadores*, or was it there to meet them? The disease in this hemisphere is kept alive between epidemics in a reservoir of rats. It passes from rat to rat by the rat louse and the rat flea, and from man to man by the human louse. Our inquiry, therefore, involves, among other things: Were the Aztecs lousy?

What was the status of rodents of the genus rat in ancient Mexico?

We are, as a matter of fact, in possession of no trustworthy accounts from which we can confidently assert that recognizable typhus epidemics occurred in Mexico before the arrival of Cortez. There is a legend, credited by Bernal Diaz and by Nicolas León, that the destruction of the Toltec city of Tollan, in 1116 A.D.,[5] was due to a typhus epidemic. This may be — but the evidence is as questionable as that concerning the nature of the plague at Athens during the Peloponnesian Wars. Fernando Ocaranza has recently reviewed the creditable records of epidemics in the Aztec kingdom, which are largely found in the Chronicles of the Franciscan Order. His evidence is helpful.

Cortez landed at Vera Cruz on the fourth of March in the year 1519.[6] In 1520, as we have said, a Negro who landed from the ship which brought the forces of Panfilo de Narvaez from Cuba came down with smallpox. The disease spread from Indian village to village, "until there was not a single healthy village in New Spain." Fifty per cent of the population died. The malady was one

[5] About the date of the arrival of the Aztecs in Mexico proper.

[6] When I entered the harbor of Vera Cruz and then proceeded, in a comfortable train, along the trail from which Cortez looked back upon his burning ships, I reflected upon the sublime valor of this man. Perhaps one of the secrets of his accomplishment lies in the fact that, unlike modern explorers, he left his wife at home. Would he have burned his ships had she been with him? No! He would have got as far as Orizaba, returned to Spain, and written a book called "Hernando and Juana Look at Mexico."

unknown to the Indians. The Franciscans thought that, had they only arrived in time, they could have arrested the outbreak by stopping the native custom of bathing when sick, through which "the blood was inflamed." Many died of hunger, there being too few unafflicted to serve the sick. The disease was called, by the survivors, "the great leprosy."

In 1531 came a second epidemic, again introduced by the conquerors, called "Tepitonzahuatl" — "the small leprosy." Many died, but not so many as in 1520. This was probably measles.

In 1545 the poor devils had another visitation. According to the Friar Geronimo de Mendieta, 150,000 Indians died in Tlascala; 100,000 in Cholula, and in other provinces numbers in proportion to the population. The symptoms were congestion (*pujamiento*), fever, bloody stools, blood from the nostrils. It might have been dysentery or typhoid fever, but the mortality is too high for these. Only plague or typhus would be likely to account for the death rate. Plague could hardly have escaped some recognizable description. Typhus, or *tabardillo,* if present, should have been recognized — for it was known in Spain at the siege of Granada, which fell on January 2, 1492. The friars knew no name for the Indian disease of 1545. But they may have been as inexperienced as many good modern doctors. It took several years before Brill's disease, seen in New York in 1906, was recognized as typhus fever — first by a Jewish physician from Poland, who happened to stroll through the wards of a New York hospital. The 1545 epidemic might have been typhus.

In 1564 the poor Aztecs were again decimated by a disease of undeterminable nature.

In 1576 a disease occurred similar to the one of 1545 — again *"pujamiento de sangre."* This was recognized as *tabardillo.* From this time on typhus epidemics were common and were definitely diagnosed. In the 1588 outbreak there was a concentration of cases in the Valley of Toluca. In this valley the natives were now mixed together, but only the Matlaxingas were severely attacked — a legend which, if true, may indicate that among the others a degree of immunity existed; and such immunity is the result of exposure to the disease, with many mild cases, in childhood — evidence of possible epidemic preëxistence of the disease among the two less afflicted tribes.

In 1595, measles, mumps, and *tabardillo* were — according to this Friar Mendieta — common among all the natives.

Mooser, who was the first to differentiate precisely between a European and a New World typhus, is inclined to believe that the disease existed in Mexico before the arrival of the Spaniards, for the following reasons. He says: "The Indians of Michoacán called typhus 'cocolixtle meco' or spotted fever: cocolixtle meaning painful fever and meco derived from 'Chichimecas,' a tribe whose members painted their bodies with red stripes and spots." Torres relates that in some parts of the state of Michoacán it was not until recent years that the name "cocolixtle meco" began to be replaced in the language of the people by the Spanish word "tifo." The Aztecs called typhus "matlazahuatl." *Matlatl* signifies net, and *zahuatl* eruption, or spots, which means an eruption arranged in the form of a net.[7] He adds that there is a hieroglyph picturing typhus in the form of a man covered with spots like a net, who is holding his head with both hands, and whose nose is bleeding. Mooser also observes that in describing an epidemic of 1573, Diaz says that "the terrible cocolixtle broke out in the surroundings of the City of Mexico," evidence that the Spanish had taken over the Indian name for the disease before they applied their own. This, we believe, is of considerable significance since, in the absence of skillful physicians, it would tend to indicate that the conquerors assumed they were witnessing an epidemic of a disease long endemic in the occupied territory, not identifying it with their own *tifo* or *tabardillo* until much later.

There is much in the historical evidence which suggests the existence of typhus fever among the South American nations in pre-Columbian days. That there were no rats in South America before the time of Blasco Nuñez, first Viceroy of Peru (1544–1546), is not a decisive argument against such an assumption. For many other rodents can harbor the typhus virus in an "inapparent" form — that is, without exhibiting obvious symptoms.[8]

The occurrence of typhus in epidemic form, at such a

[7] Ocaranza does not agree with this and quotes Robelo to the effect that none of the Mexican dialects were ever written down correctly, and that it is quite possible that the Aztec name of the disease may have been, not *matlazahuatl*, but *matlatzalatl*, meaning "ten swellings," which could possibly signify smallpox.

[8] Guinea pigs, rabbits, and a variety of mice can be infected in the laboratory — none of these animals dying as a result of the infection. Native Mexican rodents have recently been found susceptible.

period, would, however, be quite out of the question
could it be shown that the Aztecs had no lice. In our discussion of lice we have referred to the studies of Fahrenholz and particularly to those of Ewing on the varieties
of these insects found on different races of men. Ewing,
it will be remembered, found lice on the scalps of Peruvian and of Southwestern American Indian mummies. He
further mentions the presence, on the widely distributed
South American spider monkeys, of varieties of lice
sufficiently similar to those of man to arouse the speculation that the monkeys may have acquired their infesta-
tions from ancestral human forms — some tens of thousands of years ago. However this may be, the mummy
observations establish beyond peradventure that aboriginal Americans had lice of their own.

As for the actual lousiness of the Aztecs themselves,
we can find no data except Ojeda's story of the bags of
lice offered as tribute to Montezuma by the poor. But
Cowan believes that the supposed lice in the bags were
"cochineal insects," then unknown to the Spaniards. The
selling of the "long worms and lice" for food in Mexico
recounted in *Purchas's Pilgrims* should also be taken
with more than a grain of salt.

Yet more circumstantial evidence makes it almost impossible to doubt that the Aztecs had lice.[9]

The nation of the Aztecs arrived on the high plateau

[9] To be sure, Mooser writes us that in a village some distance from
Mexico City, where a recent epidemic occurred, the Indians had their
own words for other animals, but used the Spanish words "piojo,"
"caballo," for "louse" and "horse." This is important enough to warrant further investigation among other tribes.

of Mexico probably in the early twelfth century. They
came out of the Northwest from the legendary region
of Aztlan. That is about all we know, and beyond this
their origin is as vague as that of any of the other tribes
inhabiting this hemisphere before the discovery — Mayas,
Incas, Northern Indians, and Eskimos. Yet, though these
peoples, as far as we can ascertain, knew nothing of each
other, and neither had contacts nor influenced each other's
civilizations, they unquestionably come from one and the
same stock, and this is now more than conjecture — we
can reliably assert it on the basis of blood groupings.
We need not go into this technically for our present purpose. The facts are that by easily performed experiments
on the interaction of the blood serum of one individual
with the red blood cells of another, we can divide mankind into four sharply differentiated groups. Actually
there are more than these, but the four main ones will
do for the moment. The characteristics which determine
this grouping are hereditary and the inheritance follows
definite genetic laws. Consequently the study of the blood
groups has considerable anthropological-ethnological value
in revealing the relationship between different races of
men. Among Europeans, centuries of racial mixture have
obliterated origins as far as the blood groups are concerned. And the same confusion exists among Asiatics.
But among inhabitants of the Western Hemisphere, whenever reasonably pure stock has been investigated, it has
been found that a single blood group, namely that spoken
of as "Group O," predominates. Unfortunately, there
are no pure-line Incas available for study; but Mayas

have shown 97.7 per cent and the Yucatan mestizos 85 per cent Group O. A small group of Aztec descendants, obviously not pure, studied by Castaneda, were 80 per cent, and the pure-blooded American Indians are 90 per cent or over of this group. Baffin Bay Eskimos, if of pure blood, are entirely "O."

These facts have many interesting connotations, most of which have nothing to do with this discussion. The important point for us is that the similarity of blood grouping indicates the close relationship of racial stock among the inhabitants of the Western Hemisphere. And when this is considered, together with the fact that lice have been found on prehistoric mummies on at least two subdivisions of American aborigines, it appears more than likely that the Aztecs as well as the Incas were lousy.

While the historical data which we have discussed and the probability that the Aztecs were lousy combine to render likely the preëxistence of typhus on the Western Hemisphere before the Spanish conquest, it is still important to examine whether there are any facts from which we can deduce the possibility of the introduction of the disease from Europe before the earlier date of a recognizable epidemic in Mexico.

Typhus was distributed throughout Spain before Cortez landed in Mexico. If the disease was imported by the voyagers, it could not have come in infected lice. The Spanish adventurers went first to Cuba, before proceeding to points on the coast of Yucatan and Mexico. This crossing could never have been accomplished in less than several months, and the typhus-infected louse dies of the disease in at most twelve to fourteen days after feeding on infected blood. It is of course possible that the virus might have passed from sailor to sailor in a succession of typhus cases during the voyage. But had this occurred it would have been a serious matter, and it is likely that a record would have survived. In this connection there is an amusing observation of Oviedo, which we quote from Cowan. He observed that when the ships entered the tropics on their way to the Indies, the lice abandoned the sailors and attacked them again at the same point on their return. The observation is questioned by one of the supplementary writers in Cuvier's *History of the Insects*. Cowan thinks that there might be a certain amount of apparent truth in it, since heat and abundant perspiration are unfavorable to the propagation of the body louse. On the other hand, it is much more likely that in the hot weather the sailors took off their clothes, and that thereby the body louse was largely discouraged; but head lice, also capable of carrying the disease, would have remained. We have found head lice plentiful in Arabian populations in North Africa in the middle of the summer, and while not as abundant in warm countries as in the colder ones, head lice may thrive under a variety of climatic conditions. It is, for the reasons mentioned, however, unlikely that infected lice could have been transported alive on the first stages of the voyages to America.

While such transportation of the disease could, therefore, be questioned, it is not impossible that the virus may have been imported with ship rats or mice. As we have seen, the black rat has been present in Western Europe certainly since the twelfth century. It was present in France and therefore with all likelihood in Spain in the early thirteenth century, its presence in France being clearly set down in the "Roman du Renart" and in the two similar ballads, "Renart le Nouvel" and "Renart le Contrefait," which date from the late thirteenth and early fourteenth centuries. In rats, the disease can be kept going indefinitely, and may easily have survived voyages even longer than those of the Spaniards. If in this way the disease may have become endemic in Cuba, between which and many parts of Spain there was frequent communication during the early sixteenth century, it might readily have been carried from Cuba to the coast of Yucatan and Mexico. The first real epidemic in Mexico which was specifically recognized as typhus by the friars was not until 1576. Bernal Diaz, under Grijalva, left Havana on February 8, 1517, in a ship which took twenty-one days to reach the coast of Yucatan. This expedition did not proceed to Mexico proper, but went on to Florida, where half of the Spaniards were killed by the natives. Cortez left Havana on February 10, 1519, and arrived on March 12 at Tabasco, having touched Cozumel in Yucatan, and then went on to San Juan de Ulúa, or Vera Cruz, where he landed on the day before Good Friday. After that, frequent voyages were made, and it is not possible to exclude the transportation of infected rats and their distribution from the coast to the high plateaus, where the transmission of the disease to individual human beings by the first rat flea might easily have started an epidemic among a lousy population, even as it does now.

It is quite impossible to decide with certainty whether typhus was one of the gifts bestowed, with other things, by Europe upon the Western Hemisphere. But in coming to this conclusion we have learned a number of interesting facts.

CHAPTER XV

Young manhood: the period of early vigor and wild oats

I

AFTER the wars of Granada, the distribution of typhus from Spain to Italy, France, and thence northward, continued in an almost uninterrupted succession of small outbreaks; and when these were hardly spent, a new wave started from south to north after the siege of Naples in 1528. Again in 1552, true to its strategy of taking advantage of every weakness in the defenses of mankind, a serious typhus epidemic forced Charles V to abandon the siege of Metz. The investment of the city took place in the winter months, and the Imperial army, which contained Spaniards, German and Italian mercenaries, began, by early December, to suffer severely from a combination of diseases, among which were scurvy and, as usual, the enteric fevers; but the most vicious of them all was typhus. More than 10,000 men are said to have died within the month, and before the end of the year the besiegers fled, leaving the surrounding country thoroughly infected. It is perhaps at this time that the term "Morbus carcerorum" first became common for our disease, since great numbers died in the military prisons. In the villages of the countryside the pestilence did not abate until late in the following summer.

From then on typhus was never absent from the regions

invaded by returning soldiers, who lighted fuses of infection that flickered along through villages and cities wherever chance sparks lighted on inflammable material. But of these irregularly scattered and generally small outbreaks we have little precise information. It is not impossible that, without further great conflagrations, the disease might have died out in succeeding centuries had not the infection, in these early phases of its Continental life, been repeatedly renewed from the Eastern front.

The most important episode in the conquest of Europe by typhus occurred at about this time in Hungary. We have, in a preceding chapter, alluded to the belief expressed by de Toro, and repeated by Fracastorius, that typhus was imported from Cyprus; and there is much in the historical records to suggest that the evolution of the parasitism of Rickettsiæ was several hundred years more advanced in the Orient than in Europe proper. It seems hardly reasonable to accept as an accidental coincidence the fact that the two earliest epidemic waves of typhus which swept through Europe proceeded from areas in which Western armies were defending their frontiers against Oriental powers: the first during the struggle between the Spaniards and Saracens; and the second, of which we are about to speak, as a result of war with the Turks on the Hungarian front.

Since the early Middle Ages, Hungary and the Balkan Peninsula had been the frontiers of Christianity against the Crescent. In the early fifteenth century the Turks made powerful progress and, again and again, defeated Hungarian armies, making themselves masters of Serbia, not infrequently of Hungary proper, and, on occasion, investing Vienna. Eastern Hungary, for over a hundred years, was thoroughly overrun. At times it could obtain no help from the Emperors of Austria, and defended itself as best it could with the meagre forces which the King of Hungary could raise among his own people. The only frontier protection consisted of a chain of about fifty-five castles irregularly scattered along the border, without organization, fighting as much with one another as with the Turks. There was a thorough admixture of populations — the Turkish armies containing Christian captives and renegades, the so-called "matrolosos" or "quastotori"; while Turks similarly joined the Christian forces.

We have no precise knowledge of the nature of the epidemic which was Hunyadi's most powerful ally when he relieved the siege of Belgrade and defeated Mohammed II in 1456. It might have been typhus — it might have been plague. Whatever it was, the victory was a sterile one for Hungary, for the disease killed Hunyadi himself. From that time on, for a hundred years and more, epidemics, which were probably both typhus and plague, stepped on each other's heels, accompanied the incessantly warring armies, and, during the brief periods of armistice, were carried, by returning troops, to villages and towns. But it is not until 1542 that we have sufficiently precise information to permit a reliable diagnosis of typhus. In this year Joachim of Brandenburg was in Hungary with an army consisting chiefly of Germans and Italians. The disease which killed 30,000 of his men, spoken of as "Pestartige braüne," was undoubtedly typhus fever. For us, in tracing the wanderings of this infection, it is of considerable interest to know whether the Margrave's soldiers brought it with them or whether they acquired it from the Hungarians and the Turks. For, as we have seen, typhus had by this time entered Spain and Italy from the West, and was not unknown in France and Germany. A clue to the problem is furnished by an observation of great importance recorded by Györy. Györy states that the Germans suffered severely, whereas the mortality among the Hungarians and Turks was relatively slight.[1] According to contemporary observers, the mortality among the Germans was so great that a considerable part of the army never closed with the enemy, because the "Hungarian disease" killed them before the Turks had an opportunity to do so. It was for this reason that Hungary was called the "graveyard of Germans."

If this is correct, and it is not the sort of thing that would suggest itself without actual observation, it can have only one meaning: namely, that our disease was already well established in Hungary when the Imperial armies arrived. Typhus confers an immunity which, though not permanent, may still last for years, and it is commonly observed that, in endemic regions, newcomers from countries where typhus is not prevalent are much more severely attacked than are the native born. The relative immunity of the Turks and Hungarians, therefore, would tend to indicate the existence of a "herd" resistance among them, a phenomenon which could have been produced only by prolonged and constant exposure

[1] Quoted from Prinzing.

to the disease, a continuity of sporadic cases and small group outbreaks. With the returning army of the Margrave the disease was again disseminated far and wide in Europe.

The episode repeated itself on a much larger scale shortly after this (1566), when Maximilian II advanced into Hungary to protect his Eastern marches. The first passages at arms were favorable to the Emperor, who might have accomplished his purposes promptly had not typhus again taken a decisive hand. The Imperials encamped along the Danube, large bodies on the island of Komorn, on the Raab, and at Rabnitz. There was food shortage, bad water, and, Schnurrer adds with obvious horror, the beer went sour. Bad and inadequate food led to scurvy; the weather was intensely hot, dysentery and enteric fever debilitated the men; and all these things together prepared an ideal soil for typhus. Thomas Jordanus, who was present as expeditionary surgeon, has left a vivid description which makes the diagnosis unquestionable. An onset with chills was followed by abdominal pain, unquenchable thirst, delirium, and a petechial eruption which was present in almost all the cases he saw. From the army the disease spread through the surrounding country, and Maximilian was forced to abandon his campaign and make an unfavorable peace with the Turks. Eventually discipline failed and the troops scattered in bands, carrying the disease with them into Italy, Bohemia, and Germany, thence into France, through Burgundy, and northward into Belgium. Wherever these little rivulets of infection reached towns, epidemics resulted. Vienna suffered the most severe typhus outbreak of its history. Ever since that time typhus has remained endemic in Hungary, the Balkan States, and the adjoining territories of Poland and Russia. These are still, at the present day, the "home stations" from which modern European epidemics take origin.

As far as historical studies can give us a clue to such matters, we are inclined to believe that the Hungarian wars and their consequences created the circumstances which gave typhus the opportunity of passing from man to man by lice in uninterrupted cycles, short-circuiting the rat-flea phase and adapting the parasitism firmly as a man-louse-man transmission in the form which we now know as the "classical European type" or "virus humanise."

2

In describing the events which permitted typhus fever to overrun the European Continent during the seventeenth century we confine ourselves to major episodes. It would require far greater diligence than we possess — and, incidentally, would be excessively dull — were we to catalogue the almost incessant succession of minor outbreaks which harassed towns and villages during the intervals between great epidemics. Once thoroughly established west of the Balkans by the circumstances described in the preceding section, typhus began to spread in all directions, not unlike a brush fire, now low and smouldering and, perhaps, in places almost extinguished; again slowly burning its way into new regions; at all times ready to burst into destructive flame when fuel was available. In this it was not alone during that century which was, of all periods of so-called Christian civilization, the most miserable for man. Plague, which was then the inseparable, ferocious companion of typhus, had never been entirely extinguished since the fourteenth century; smallpox, diphtheria, the enteric fevers, and all the lesser scourges were constantly alert; and the chronicles of the years are pitiful records of famine, pestilence, and unbelievably savage wars.

The wretchedness of the period is vividly set forth in the account in which Lammert has compiled, year by year, from 1600 to the end of the Thirty Years' War, the dreadful companionships of pestilence and warfare. Lammert was a *Bezirksarzt* (district physician) at Regensburg who studied the local chronicles of different regions in Germany. He had the quaint habit of heading the sections treating of successive years with records of weather conditions, crop reports, and, invariably, with statements concerning the quality of the wine.[2] Thus, in 1602, we find:

[2] Lammert's preoccupation with the weather is quite natural. Earlier books on epidemic disease are dominated by the idea that natural phenomena, volcanic eruptions, earthquakes, abnormal weather conditions, eclipses, and so on, were largely responsible for epidemics. Modern epidemiology recognizes that atmospheric conditions, temperature, humidity, have distinct effects upon the occurrence and spread of disease, facts for which there are rational explanations. Lammert's attention to the vintage is not so readily explained, but may not be as illogical as it seems at first sight. The habit of wine drinking may well have had its origin in a crude public-health conception. Wherever, in the Middle Ages, people lived together in groups, the water was contaminated. Men knew by experience that drinking water was apt to be dangerous. There is a passage somewhere in Froissart which tells us that an army marching into Spain was rendered helpless by an outbreak

"There was a severe winter, a cold April, a hailstorm in the summer. The wine was scarce and of poor quality. In this year there was plague in the Palatinate, through Saxony and Prussia. In Danzig 12,000 people died in one week. There was a smallpox epidemic in Bohemia; another in Silesia. In Southern Germany there raged the terrible *Bauchkrankheit* [probably dysentery or typhoid]. There was a famine in Russia accompanied by pestilences of plague and typhus, and in Moscow alone [probably a gross exaggeration] 127,000 people are said to have died of pestilence."

Each year repeats the grim story. We choose another at random. Thus: "In 1613, when the wine was plentiful but sour, the Hungarian disease [typhus] swept across Württemberg and the Tyrol. *Hauptweh* [typhus] reigned in Magdeburg. There was plague in Regensburg, in Leipzig, in Bohemia and in Austria, whence it spread eastward." Such is the story year by year until 1618, when the Thirty Years' War began.

The Thirty Years' War was the most gigantic natural experiment in epidemiology to which mankind has ever been subjected.[3] Europe, as we have seen, was a spot map

---

of dysentery which occurred because the wine gave out and the men had to drink water. This was an army of 20,000 men, and the implication is that the entire 20,000 drank no water until they were unable to get hold of wine.

[3] There is a relatively new method of investigating infectious disease which is called "experimental epidemiology." It consists in setting up large colonies of mice, rats, guinea pigs, rabbits, or other animals susceptible to spontaneous infection with some microörganism, and then introducing, into such a colony, under a variety of controlled conditions, one or more infected individuals. In this way the circumstances which

of constant small outbreaks of every conceivable infectious disease; and through this area, for a little over twenty-nine years, armies marched and countermarched, and disbanded soldiers, fugitives, and deserters vagabonded far and wide. Famines resulted and populations wandered in fugitive hordes toward food and protection. Wherever men traveled, disease followed them.

The history of these epidemics can be fully understood only against the background of the conditions which gave rise to them; and a conception of these conditions can be best conveyed by episodes such as those taken by Lammert from contemporary records. There is an embarrassment of choice. The following random example is translated verbatim from a paragraph included by Lammert in his account of the year 1632. We might, with equal illustrative value, have chosen almost any other year: "When Gustavus Adolphus, after taking Memmingen, prepared to overrun Southern Germany, he was held back by the news of Wallenstein's triumphal progress in Saxony. Memmingen was soon recaptured by the Imperial army. The former 'Reichstadt,' Kempten, fell into

favor spread can be observed and much information obtained. The method has proved useful, but has its inevitable limitations, because a mouse or guinea-pig colony in a closed compartment can never entirely simulate the complex conditions of human association. Nature sets up her experiments of epidemiology in times of war and famine, and when, as in the wars of the late nineteenth and the twentieth century, these dreadful experiments can be observed by a competent medical profession, much of value to mankind may be learned. It can well be said that nobody won the last war except the medical sciences. The profit was not worth the loss, but the increase in sanitary and medical knowledge was the sole determinable gain for mankind in an otherwise utterly disastrous catastrophe.

the hands of the Swedes, and the contemporary chronicle written by Dr. Ph. Jak. Karrer records the revolting occurrences in this town." The good Lammert writes, *"Die Feder sträubt sich,"* the pen revolts against recording "such bestialization of man." When women were captured, their breasts were cut off; mothers, with their children and servants, were thrown into the river. The soldiers killed the local surgeon, ravished his daughter, gouged out her eyes, and threw her out of the window together with her dead father. In the presence of husbands and parents, later to be murdered, wives and daughters were raped. Finding a housewife standing before a kettle of boiling water, the Swedes cut off her hands, dipped her head repeatedly into the kettle, and decapitated her. Six little children were found murdered in a cellar. On the thirteenth of January the city again fell into the hands of the Imperial troops. The atrocities which the conquerors now perpetrated upon what was left of the population, recorded by Dr. Gabriel Furtenbach in what he appropriately calls his *Jammerchronik*, defy all imagination. This happened shortly before the march of Gustavus Adolphus on Nuremberg, where typhus wrought appropriate vengeance on both armies.

Prinzing divides the epidemiological history of the Thirty Years' War into two main periods: the earlier, from 1618 to 1630, when typhus was the chief scourge; and the later, from 1630 to 1648, when plague gained the ascendancy. It must not be forgotten, however, that throughout the entire period both diseases raged together and were sturdily reënforced by dysentery, typhoid

fever, diphtheria, smallpox, scarlet fever, and a variety
of less deadly confederates.

The very beginning of the war was accompanied by a
severe typhus epidemic. The army of Mansfeld, after
the battle of Weissenburg, marched through the Palatinate into Alsace, and everywhere left typhus behind it.
This started a succession of epidemics through Bohemia
and the South of Germany. Thence the disease was carried into the North with the troops of Wallenstein and
Tilly, where in 1625 plague and typhus reached their
culmination. Devastation of the fields drove peasants into
the cities, and the pestilence spread into Strassburg, Mann-
heim, Frankfort, Mainz, Nuremberg, and all the smaller
towns. In Metz, typhus again appeared in 1625, and
then spread through Verdun into France. Saxony suffered
severely from typhus and plague after the battle of Brei-
tenfeld in 1631. Plague now gained the ascendancy and
the two diseases together traveled with the rapidly moving
armies, remaining behind when the soldiers departed,
and spreading from innumerable foci into the surrounding country. Bavaria was almost depopulated at this time.

In June of 1632 Gustavus Adolphus besieged Nuremberg. An enormous number of fugitives and troops had
congregated in the city. After eleven weeks of stubborn
resistance food and supplies gave out. The Hungarian
disease (typhus) and scurvy spread among besieged and
besiegers alike. In the town some five thousand victims
are inscribed in the church records, and these are only a
fraction of the dead. The nun, Maria Anna Junius of
Bamberg, writes in her chronicle under the date of No-

vember of this year: *"War damals grosse Theurung und Sterb zu Nürnberg, dass in 7 Wochen 29,000 Menschen gestorben."*

The Swedish army suffered no less. Hunger and disease destroyed all discipline, and the impoverished peasants of the surrounding country fell prey to the ferocity of the soldiers. After a final unsuccessful attempt to storm the town on September 3, the Swedish King was forced to retreat. He left desolation behind him: fields were devastated; villages were heaps of ashes, their streets foul with the stench of dead bodies; in one district only a quarter of the original population survived; many of the few survivors, citizens, peasants, and stray soldiers, contracted disease by invading the abandoned encampments of both the Swedish and the Imperial troops in search of food and plunder. Typhus and plague were again scattered far and wide. Typhus had raised the siege and had forced both armies to retreat without battle.

The epidemic disasters of the Thirty Years' War, however, were not limited to the actual scenes of struggle. Infection was constantly carried across national borders. In 1624 over ten thousand people died in Amsterdam. France was invaded by typhus at almost the same time. Western Provence was, at this time, the scene of the ferocious war waged against the Calvinists. Montpellier was besieged in 1623, and a disease broke out which is described by Lazarus Riverius as "febris maligna pestilens." His description, which Murchison cites in detail, is unmistakably that of typhus. "The skin was marked by an eruption of red, livid or black spots resembling flea bites, which appeared from the fourth to the ninth day, over all parts of the body, but most frequently on the loins, chest and neck." The infection remained in the district and again became epidemic in 1641. From Montpellier typhus, together with bubonic plague, spread northward. In 1628 (we take our figures from Prinzing), there were 60,000 deaths in Lyons and 25,000 in Limoges. It extended to Paris and Avignon, toward the Pyrenees and along the Mediterranean littoral.

When the Thirty Years' War was ended, no corner of the European Continent was left without its foci of infection. And although the dreadful period of this war overshadows all other events of the century, the subsequent years were by no means peaceful ones. The campaigns of Turenne, the wars in the Netherlands and in Russia, and continued warfare with the Turks, — especially the siege of Vienna in 1683, — offered typhus all the opportunities it needed for continuous activity. And in Italy — especially Sicily — famines gave the disease a free hand in some of the most severe epidemics of its history. Meanwhile France itself was not spared, and 1651 and 1666 were calamitous typhus years for Poitou and Burgundy.

On the Eastern battlefields, where the struggles between Russia, Austria, and Hungary continued without respite until well into the eighteenth century, the disease became more and more firmly implanted, leading to the establishment of the permanent foci of which we have spoken.

3

In the early epidemiological records of England there is no evidence that typhus fever existed before it had become firmly established on the Continent. There were, of course, many dreadful epidemics — such as the "Drif" or "famine fever" of 1087, mentioned in the Anglo-Saxon chronicles: "A.D. 1087 after the Birth of our Lord and Saviour Christ, one thousand and eighty-seven winters; in the one and twentieth year after William began to govern and direct England, as God granted him, was a very heavy and pestilent season in this land. Such a sickness came on men that full nigh every other man was in the worst disorder, that is in the diarrhœa; and that so dreadfully, that many men died in the disorder." This was quite evidently not typhus — possibly dysentery and enteric fever combined with the deficiency diseases incident to famine. We are equally in the dark concerning the nature of the famine fevers of 1196 (described by William of Newburgh), of 1258, and of 1315. Lieutenant Colonel W. P. MacArthur, who has written a scholarly review of typhus in ancient England, is inclined to believe that these epidemics, as well as the diseases associated with gaols in London in 1414, were, in part, typhus. But he suggests this only in view of the circumstances under which the outbreaks occurred, and admits the complete lack of basis for any specific diagnosis in the very vague descriptions. In view of the apparent absence of epidemic typhus from Europe before the fifteenth century, it would seem far more likely that the disease, once well established on the Continent by the middle of the sixteenth century, had then crossed the Channel and the Irish Sea, where it found a fertile soil in the crowded and filthy villages and towns inhabited by thoroughly lousy populations.

In England some of the earliest unmistakable ravages of our disease were in the prisons, where it became known as the dreaded "gaol fever" or "jayl fever." MacArthur tells us that the English prison system was "thoroughly rotten from top to bottom. . . . Some gaols were private property, rented by the gaolers, who reimbursed themselves by fees exacted from the prisoners and their friends. . . . Prisoners were loaded with chains so that gaolers could extort bribes for 'easement of irons.' . . . Prisons were scandalously overcrowded and indescribably filthy." These conditions continued for centuries, until after 1770, when John Howard, the first great advocate of prison reform (who himself died of typhus as a result of his tours of inspection), wrote his pamphlet on *The State of the Prisons in England and Wales*. Typhus flourished in the gaols and, on occasion, escaped and ran riot in the surrounding country. This it did in particularly dramatic fashion in what are known as the Black Assizes. There were a number of these: at Oxford in 1577; at Exeter twelve years later, and, the last serious one, at the Old Bailey in 1750. The following facts are largely taken from MacArthur.

In 1577 there was committed to prison at Oxford a certain Rowland Jencks, a Catholic bookbinder who was accused of speaking evil of "that government now settled," of profaning God's Word, abusing the ministers, and staying away from church. Considering the times, he appears to have been a fellow of spirit and conviction. Just before his trial started a number of inmates of the prison at Oxford had died in their chains. The trial, at which Jencks was condemned to have his ears cut off, took place in a court unusually crowded because of the lively public interest aroused by the Jencks case. Soon after the trial typhus began to appear among those who had been present. MacArthur tells us that Sir Robert Bell, the Lord Chief Baron, and Sir Nicholas Barham both died, as did the sheriff, the undersheriff, and all of the members of the Grand Jury except one or two. The total deaths were over five hundred, of which one hundred were members of the University. The occurrence created considerable excitement, and even Sir Francis Bacon took the trouble to investigate, attributing the disease to the stinks that "have some similitude with man's body and so insinuate themselves." [4] The theories of the day attributed most of these mysterious infections to vitiated air, a not unnatural assumption under the circumstances. In this particular case papistical evil magic was suspected in the form of winds compounded in Catholic Louvain and secretly let loose at Oxford, *diabolicis et papisticis flatibus*. Jencks himself, MacArthur says, though deprived of his ears, escaped the infection, settled in Douai, where he obtained employment as a baker in the English College of Seculars, and lived thirty-three years after the

[4] In the medical jargon of to-day these would be known as "homologous stinks."

disastrous Assizes. Reasoning from the manner of spread of the disease among the learned audience, MacArthur reaches the conclusion — in which facts force us reluctantly to concur — that no inconsiderable number of the faculty of Oxford College were, at this time, lousy.

The Exeter Assizes were, in a general way, similar in circumstance to those which shortly before had occurred at Oxford. That the condition of jails nevertheless continued unchanged is witnessed by the Old Bailey outbreak which came two centuries later (1750) and was investigated and described with accuracy by Sir John Pringle, Physician in Chief to His Majesty's Forces and later President of the Royal Society.

In England, generally, typhus penetrated all corners of the Island. The description by Thomas Willis, the Oxford anatomist, leaves no room for doubt that the disease which decimated both the Parliamentary and the Royal armies at the siege of Reading in 1643 was typhus (Murchison). And in 1650 an epidemic of the same character "converted the whole Island into one vast hospital." And, just as on the Continent typhus and plague marched hand in hand at this time, the Great Plague was accompanied by typhus which preceded the accumulation of plague cases during the cold winter of 1665.

Exactly when typhus reached Ireland, which later became and still remains one of the most impregnable strongholds of the disease, is uncertain. Murchison says that the first precisely recorded epidemic was that observed at Cork in 1708, but there is reason to believe that, as the "Irish Ague," it had existed long before that time.

CHAPTER XVI

Appraisal of a contemporary and prospects of future education and discipline

I

WERE we engaged in writing medical history instead of biography, it would now be our task to describe, chronologically and geographically, the almost uninterrupted succession of typhus epidemics which spared no byway and corner of Europe throughout the eighteenth and a large part of the nineteenth century. Such records, however, though indispensable to the student of infectious diseases, would contribute little to our present purpose of setting forth the character and habits of the subject of our biography. They are available, moreover, in forms far more scholarly and thorough than anything we could achieve, in the treatises of Ozanam, Hirsch, Haeser, Prinzing, and others, from all of whom we have freely borrowed. The specialist, in studying the epidemiological data of former times, not infrequently finds observations and information which, in the light of modern knowledge, become valuable clues to unsolved problems. From the biographical point of view, however, circumstantial accounts of the typhus outbreaks, of which no decade of the period of which we speak was entirely free, would be dull with constant repetition. The circumstances of occurrence, sequence of events, and manner of spread were always the same in principle. Typhus had come to be the inevitable and expected companion of war and revolution; no encampment, no campaigning army, and no besieged city escaped it. It added to the terror of famines and floods; it stalked stealthily through the wretched quarters of the poor in cities and villages; it flourished in prisons and even went to sea in ships. And whenever circumstances were favorable it spread through countries and across national boundaries. If there were any significant differences between the eighteenth-century manifestations of typhus and those of preceding periods, they consisted in the fact that, in addition to the major epidemics that regularly accompanied human strife and misfortune, there were now numerous smaller group outbreaks, scattered here and there in widely separated regions; and on the Eastern frontiers, possibly in Italy, Spain, and parts of Germany as well, the infection was sporadically present at all times, much as typhoid fever is with us now. The disease had now become widely disseminated and, in areas where circumstances were favorable for slow propagation, firmly implanted.

As a matter of fact, until the last decade of the nineteenth century mankind changed very little as concerns those customs and personal habits which determine its relationship with typhus fever. The extraordinary political, philosophical, and scientific awakenings which shed so much lustre over the eighteenth century found no reflection in that fastidiousness of physical living which alone can curtail the homicidal aggressiveness of our disease. Elegance of manners and dress was never more assiduously cultivated, but cleanliness did not keep pace.

Even a superficial survey of the evolution of human cleanliness — a subject which well merits a far more thorough treatment than we can give it here — reveals that its development has lagged far behind the intellectual, æsthetic, and moral progress of man. Cleanliness was not akin to intelligence and certainly not akin to godliness; we have seen many godly people who — However, one must not take these old adages too seriously. This one — like "Honesty is the best policy," "Virtue is its own reward," "Waste not, want not," and so forth — merely expresses the cherished wish of those who dream of unattainable perfections. In a perfect world cleanliness would be at least akin to intelligence, and virtue would be its own reward. These proverbs are of the same order of thought as Keats's "Beauty is truth," a postulate about which — in spite of his inexperience of the world — his short service as a medical student might have enlightened him.

However, we have wandered from our theme, which was that the development of cleanliness lags far behind the progress of intellectual and æsthetic attainments. Indeed, observation, especially of some of our artistic contemporaries, has often led us to speculate whether there might not be something mutually exclusive in the two tendencies. At any rate, in spite of the extraordinary enrichment of mankind in other blessings of civilization during the two brilliant centuries of which we speak, cleanliness did not make headway until medicine had begun to establish the physical perils of filth on a scientif-

ically demonstrable basis. Thus we learn of the education of a princess (of about 1700) that *"on lui apprit à se
décrotter les pieds . . . pour ne pas polluer sa couche.
. . . Elle savait que lorsqu'on se presse la narine en souf-
flant, il faut incontinent marcher sur ce qui tombe à terre."*
. . . Or *"que c'est chose vilaine . . . de prendre au col
les poux, puces et autres vermines pour les tuer devant les
gens, à moins qu'on ne soit dans l'intimité."*

The new freedom which was preached by Voltaire and
Rousseau did not include freedom from vermin. The purpose of wigs worn on shaven heads has been dealt with
elsewhere. Cities and villages stank to heaven. The streets
were the receptacles of refuse, human and otherwise. The
triangular intervals which one sees between adjacent
mediæval houses in streets still inhabited are apertures
through which waste, *pots de chambre,* and so forth, could
be conveniently disposed of from the upper stories. The
opulent used the *chaises percées* as the last word in fastidiousness. Baths were therapeutic procedures not to be
recklessly prescribed after October. The first bathtubs
did not reach America — we believe — until about 1840.
And public bath houses lacking sanitary laundry arrangements were as likely to spread disease as to arrest it.
Schools, prisons, and public meeting places of all kinds
were utterly without provisions which might have limited
the transmission of infection. When the windmill ventilation device was installed on Newgate in 1752, MacArthur
says that it was "rumored" that two men fell dead when
the first blasts from the exhaust pipe struck them. This
is probably, as MacArthur says, an exaggeration, but

even the false rumor conveys some idea of the probable
condition within the building.

2

Considering these circumstances, it is not surprising
that typhus fever ran riot through Europe and, occasionally, reached America during the period of which we
write. The turbulent events of the eighteenth century
had carried the infection into the remotest corners of the
civilized world. No longer was it necessary to seek the
origins of renewed outbursts in the East, though continuing wars with the Turks undoubtedly added occasional
sparks. The wars of the Spanish, Polish, and Austrian
Successions, all of which occurred in the first half of the
eighteenth century, provided the old opportunities never
overlooked by typhus. In all of them, pestilences, some of
which have been discussed in preceding chapters, started in
the armies, spread through Central Europe. At the siege
of Prague alone, 30,000 people — including all the
French medical staff — died. Another wave, during this
same period, swept through Scandinavia, probably via Russia, and crossed into Germany. A little later it appeared
with deadly violence in Paris and spread into the provinces.
Its presence in Ireland was first reliably reported early
in the century by O'Connel, and it was widely epidemic
by 1718. As the "Irish Ague" it probably occurred there
much earlier — but this cannot be positively determined.
In 1720 famine gave it its opening at Messina; a disastrous outbreak occurred in Moscow in 1735; and in
1740, after a decade of relative quiescence, it suddenly

reappeared — almost simultaneously and with renewed vigor — in Central Germany and in Ireland. In Ireland the occasion was the potato famine of 1740. It is noticeable that in this century, with the development of industry, decline of trade and unemployment began to play into the hands of typhus fever — together with wars and agricultural disasters. There were severe outbreaks in connection with difficulties in the textile industry in Flanders and in Austria, a demonstration of its relationship with purely economic hardship.

From now on typhus again followed the armies. It campaigned with the British in Flanders, after Dettingen (1743), and again in the Spanish wars in 1762. In the same year it lighted up in Italy, where, abetted by famine, it lingered, rising and falling, until 1769. The Naples epidemic of 1764, described by Fasano, was the most dreadful episode of this era. Speaking of the outbreak, Haeser makes the illuminating remark that mortalities were lowest wherever there was a shortage of doctors, a circumstance quite probably true, since the medical conventions of the day favored copious bleeding.

The Seven Years' War, the French Revolution, and the Napoleonic campaigns in Europe and in Spain were all more destructive of life by the activity of our disease than by the power of cannon, rifle, and bayonet. Toward the end of the eighteenth century and the beginnings of the nineteenth, England, which had been relatively spared by typhus during the Continental wars, was seriously invaded. As the Continental epidemics began to decline, toward 1798, the infection reëntered England, probably from Ireland, where poor crops and famine had again given our disease its opening. The succeeding two decades were typhus years in both islands. The disease reached its culmination in 1816 to 1819. During the great Irish epidemic of these years it is recorded that there were no less than 700,000 cases among the 6,000,000 inhabitants. At almost the same time (1818) Italy was the scene of another wave of infection, which swept southward from the Alps to Sicily.

"Ship fever" was one of the common popular designations of typhus throughout the eighteenth century. Next to battle casualties and scurvy, it was the most dreadful affliction of navies. Lind was one of that extraordinary group of physicians which the eighteenth century produced in all countries of Europe, who reasoned correctly from circumstantial evidence and predicted from pure clinical observations a great many things which were later substantiated by scientific investigation. He was physician to His Majesty's Hospital at Haslar, near Portsmouth, and left two papers on fevers and infection, an essay on the most effectual means of preserving the health of seamen, and a small volume on diseases in hot climates. Among other things, recognizing — as many others did at this time — the great importance of fruit, greens, and vegetables for maintaining health on long voyages, he developed ingenious methods for the preservation of orange and lemon juice and of vegetables. The fruit juices were kept from deteriorating by putting them into small pint bottles and covering the surface with a layer of olive oil before tightly corking them. Leeks and other vegetables he cut into short lengths, and sprinkled them with a thin layer of dried bay salt, packing the entire mass of vegetables in salt. When the salt was washed out as much as three months later, the preserves could be prepared as fresh vegetables and had apparently retained the properties for which they were valued. His views on the effects of wine and stronger drinks such as "garlic brandy" are perhaps not so medically sound, but may have contributed considerably to his popularity in the navy. In connection with typhus, his notable contribution consists in the description of the disease as one of the most disabling scourges of the royal navy, with its dissemination from the ships to the hospitals on land, and thence to the surrounding country.

There was at this time in England a lively controversy concerning the importance of ventilation. In spite of the popular belief in the dangers of contaminated air, Lind was quite sure that ventilation and the supply of clean air had very little effect on the spread of disease. As far as typhus fever itself was concerned, he was quite convinced that the infection was carried not only on the bodies of men, but upon clothes, on all kinds of material, — wool, cotton, linen, — and might cling for some time to wooden beams, chairs, bedsteads, and such. He cites, in defense of his views, many observations, among which is the death of seventeen of twenty-three people who had been employed in refitting old tents in which patients had been cared for. He speaks of the infection of the sleeping quarters in ships, and advocates fumigation. The materials used for disinfection were probably not very effective. They consisted of the burning of tobacco, steam from charcoal fires, the evaporation of camphorated vinegar, and the smoke from pitch tar and gunpowder. However, combined with these ineffective methods of fumigation, Lind ordered thorough scouring and cleansing and the removal of bedding and all clothing to the decks, for sun and air. Likewise, he recommended that physicians and nurses change their clothing when leaving the hospital. Altogether the measures advocated by Lind — without his having any suspicion of insect transmission — must have saved a considerable number of lives.

3

The last half of the nineteenth century marks a turning point in the epidemic history of the Western World. Transmissible diseases were, of course, still plentiful; and scarlet fever, diphtheria, meningitis, and measles — which had been previously masked to some extent by the more rapidly spreading and violent contagions — now attained greater prominence. Cholera also had penetrated into Europe on several occasions during this period. But except for influenza, the pestilences which had, throughout preceding centuries, caused the most widespread destruction were distinctly declining and were becoming more limited in regional distribution. Plague had practically disappeared. Smallpox, which, after almost complete conquest by Jennerian vaccination, burst into renewed energy in the thirties, had again to be brought under reasonable control by the practice of revaccination. This practice was introduced in 1823 and widely applied before

1850. Typhus was becoming more and more rare and was limited to restricted areas on the Eastern frontiers and in Ireland — except for the occasional epidemic recrudescences which, following wars and periods of economic depression, proved that the seeds of the disease had not been entirely stamped out. It reached the United States early in the century — probably in the imported form, since it remained limited to cities of the Eastern coast. The Philadelphia outbreak of 1837 was the one during which Gerhardt and Pennock made their valuable contributions to differential diagnosis. The outbreak in Silesia in 1846 and that in London in 1862 were the direct consequences of industrial depression. In Silesia — always in contact with the endemic centres of the East — the collapse of the textile industry was responsible. In England, according to Murchison, the epidemic was the result of the great crowds of unemployed that wandered into the cities. Here, also, we may assume that infection may have been reintroduced some six years before, with soldiers returning from the Crimea.

During the Civil War — in which, in the Federal armies, 44,238 were killed in battle, 49,205 died of wounds, and 186,216 died of disease — typhus was not very important. And in the relatively short European wars, the French campaign in Italy, the Austro-German and the Franco-Prussian wars, typhus played a negligible rôle. It is of considerable interest, in anticipation of what we shall have to say of typhus and the World War, that in the Franco-Prussian struggle of 1870 there was little or no typhus in either of the contending armies, except for a moderate number of cases (252) among the Algerian troops; furthermore, there is considerable question whether the disease occurred in any of the besieged cities. At the same time Prussian troops on the Russian border were never entirely free from the disease. Smallpox, dysentery, and typhoid fever had now taken the places of plague and typhus as the major scourges of armies.

It is not easy to account for the decline of great epidemics in Europe after 1850. One might assume an unaccountable cyclic change in the characters of prevalent diseases. On the other hand one is inclined to give much credit to the coöperative forces of modern civilized society when one considers the immediate calamitous consequences which followed the temporary return to quasi-mediæval conditions in Russia and the Near East during and after the last war. These forces were manifold and it is impossible to give any one of them the first place. Of considerable importance, no doubt, is the fact that wars, during this period, were of short duration and operations were within relatively circumscribed areas. Another factor, not to be underestimated, was the safeguard against famine provided by the development of intensive agriculture and the perfection of railroad transportation, which prevented the former prolonged isolation of famine districts from supplies of food and succor. Of at least equal importance was the rise of modern medicine, the development of methods of diagnosis, rational approaches to prevention, and the organization of local, national, and military health supervision which gradually extended into all ramifications of community life. To describe these in anything like completeness would require another, perhaps useful, but exceedingly dull volume.

It is a curious and heartening fact that international coöperation in the prevention of epidemics placidly continues, however hostile or competitive other relationships may become. At the present moment, — while the world is an armed camp of suspicion and hatred, and nations are doing their best, by hook and crook, to push each other out of the world markets, to foment revolutions and steal each other's political and military secrets, — organized government agencies are exchanging information concerning epidemic diseases; sanitarians, bacteriologists, epidemiologists, and health administrators are coöperating, consulting each other, and freely interchanging views, materials, and methods, from Russia to South America, from Scandinavia to the tropics. It is perhaps not generally known that for several years, during the most turbulent period of the Russian Revolution, the only official relationship which existed between that unfortunate country and the rest of Europe consisted in the interchange of information bearing on the prevention of epidemic disease, arranged in coöperation by the Health Commission of the League of Nations and the Soviet government.

It is all a part of the strange contradictions between idealism and savagery that characterize the most curious of all mammals. It leads to the extraordinary practice of what is spoken of as "saving at the spigot and wasting at the bung."

Thus, during the decade immediately preceding the

World War, typhus fever was leading the quiet bourgeois existence of a reasonably domesticated disease. It was, to be sure, causing its periodic localized epidemics in China and in Mexico, was sporadically occurring in North Africa and the Near East, and was continuing (with a declining rate) in Ireland, where there were only seventy deaths between 1899 and 1913, although the "Green Island" was regarded as the only Western country with any considerable typhus incidence. In American cities the infection was present in a mild form, as Brill's disease (about 528 cases in New York and Boston from 1900 to 1930), and undoubtedly it was occurring in the same relatively tame manner in many other parts of the world, in South America, in the Mediterranean basin, and in remote parts of the Orient, where — though unsuspected at that time — it has now been detected. However, there were no great epidemics, and the only countries in the world where there were a sufficient number of annual cases and deaths to justify their designation as "endemic centres" were Russia, Poland, and parts of Eastern Austria (Galicia).

In these regions, as well as in the adjacent Hungarian and Balkan territory, typhus kept claiming its annual toll of victims — though epidemic dimensions were rarely approached except in the presence of the circumstances of famine or war. Thus cases in Russia usually averaged about 90,000 a year: the lowest, 36,887, in 1897; the highest, 184,000, in 1892, when there was a famine. In the Balkan countries morbidity rates increased during the war years, 1912–1913; but even then no true epidemic occurred. Western Europe was practically exempt. The organization of modern life and the forces which we have enumerated in a preceding paragraph were holding typhus to an armed truce. And then, for the first time in the ages-old struggle between the two enemies, the strategic initiative passed into the hands of man, with the discovery, in 1909, by Charles Nicolle (to whom we have dedicated this book), of the louse transmission of typhus fever from man to man. For the first time in all the centuries of a one-sided warfare, with man forever in the open and typhus ever in ambush, the victim was in a position to organize a rationally planned and strategically sound defense against his historic enemy.

If warriors and politicians and patriots and all the other people responsible for wars had only left the world alone for another hundred years this discovery might, without further scientific advances, have sounded the knell of epidemic typhus in the West.

But then a Grand Duke was murdered at Serajevo and everybody lost their heads, ourselves and T. Roosevelt included — except Mr. Wilson, who lost his two years later; and the bands played the "Wacht am Rhein" and the "Marseilleise" and "God Save the King" and "Gott erhalte Franz den Kaiser" and "Boje tsaria Khrani" and "Ustaj, ustaj, Srbine" and, several years later, the "Star-Spangled Banner." And the barbed-wire kings and the T. N. T., corned beef, and ordnance people, and the ship jobbers and the shoe manufacturers and the khaki pants trade, and so forth, and so forth, laid the foundation for a new and Hollywoodian aristocracy that lasted until 1929. And God was on everyone's side. And when we had all gone to war and the stage was set, typhus woke up again.

Not everyone realizes that typhus has at least as just a reason to claim that it "won the war" as any of the contending nations. Many a French barroom fight might have been avoided if this had been clearly understood.

4

It raised its ugly head first in Serbia. This valiant little nation had hardly recovered from the Balkan troubles when, in July 1914, Austria declared war and immediately attacked. Belgrade was bombarded and the Serbian government retired to Nish. The terrified villagers of the border regions began to move southward toward safety with all their portable possessions. Early efforts of the Austrians to cross the Sava, near Belgrade, were repulsed. But later, attacking from the Bosnian border, they succeeded, in November (not without reverses in which 20,000 Austrian prisoners were taken), in capturing Valjevo and Belgrade. On December 2 the Serbian army counterattacked and the Austrians were driven back across the Drina and the Sava, and Valjevo and Belgrade were retaken. As a result of these battles Northern Serbia was a shambles. Villages were in ruins and the noncombatant population was crowding its way toward the South.

Typhus began to show itself in the Serbian army in November. It is probable that it occurred, at the same time, among the invaders. In addition to their own

troubles the Serbs now had about 60,000 to 70,000 prisoners on their hands, some of them sick and wounded.
They were short of shelter for their own dispossessed
civilian population; there were no adequate quarters for
their prisoners. Most of their able-bodied adults were
with the colors. There were less than four hundred doctors in the country, almost all of whom sooner or later
contracted the disease, 126 of them fatally. The few
existing hospitals were soon overflowing, and others had
to be improvised in buildings which often lacked sanitary
provisions of all but the most primitive order. There were
practically no nurses. There were no beds, no linen, no
medicines. Eventually there were hardly enough grave
diggers. It is impossible to state, with any accuracy, just
where the epidemic started. The first accumulation of
cases occurred among Austrian prisoners at Valjevo. Dissemination to all parts of the country was almost immediate. The infection traveled with the wandering population, with prison trains, and with moving troops. Through
February and March the epidemic flared up with a
speed and violence never equaled in any typhus outbreak of which we have reliable record. In April — when
it reached its culmination — the new cases per day ran
into many thousands. For a time 2500 were daily admitted to the military hospitals alone. The mortality
ranged from approximately 20 per cent during the rise
and decline to 60 and even 70 per cent at the height of
the epidemic. In less than six months over 150,000 people
died of typhus. Not less than one half of the 60,000 Austrian prisoners succumbed.

During all this time Serbia was practically helpless. Yet Austria did not attack. Military operations were largely confined to a short bombardment of the railroad station in Belgrade at about four o'clock in the afternoon, during which everyone stayed away from the trains. Austrian strategists knew better than to enter Serbia at this time. The probable results were obvious. Typhus — while scourging the Serbian population — was holding the border. The Central Powers lost six months during the most critical time of the war. It is anybody's guess as to the effect which this delay may have had on the early Russian and even on the Western campaigns. It is at least not unreasonable to believe that a quick thrust through Serbia at this time, — with its reactions on Turkey, Bulgaria, and Greece, — the closing of Salonika, and the establishment of a Southwestern front against Russia might have tipped the balance in favor of the then very vigorous Central Powers. Typhus may not have won the war — but it certainly helped.

Typhus from now on took over its historic rôle along the entire Eastern front. It flourished as usual in all the Eastern armies, but was kept, by extraordinarily effective sanitary measures, — bathing and delousing, — within reasonable bounds among the Austrians and Germans. Though it penetrated into the prison camps in Central Europe, it was successfully prevented from spreading to the civilian populations. Among the most remarkable phenomena of the war is the total absence of typhus from the Western front. No completely satisfactory explanation for this can be offered. Soldiers in the trenches on this front were as universally lousy as soldiers have always been. And a louse-borne disease, Trench fever, closely allied to typhus, was common. We can attribute it only to the fact that the armies were — on both sides — more afraid of typhus than they were of shot and shell. The Central Powers, realizing that a typhus epidemic, introduced with troops transferred from the East, would lose them the war, took the utmost precautions to avoid this. And army sanitary organizations, in all the forces, were ever conscious of the possible peril, alert for suspicious cases, and usually quick to resort to wholesale delousing. The mortality of lice in this war must have been the greatest in the history of the world.

In Russia alone did typhus attain its mediæval ascendancy. During the first year of the war only about 100,000 cases occurred in Russia. After the retreat of 1916 the recorded number rose to 154,000. From then on, for obvious reasons, figures are unreliable, but there is no question that the disease increased steadily and rapidly. Revolution, famine, epidemics of cholera, typhoid, and dysentery, helped. There are no words to record the dreadful sufferings of the Russian people from 1917 to 1921. We are concerned with typhus alone. And from the careful and conservative calculations of Tarassewitch, it is likely that, during these years, there were no less, and probably were more than twenty-five million cases of typhus in the territories controlled by the Soviet Republic, with from two and one-half to three million deaths.

We have said nothing of the epidemics in Poland, Rumania, Lithuania, and the Near East, but we are —

and the reader surely is — weary with horrors. Moreover figures, when they begin to approximate those of President Roosevelt's expenditures, begin to anæsthetize the mind and lose effect.

The typhus records of the World War are reassuring as far as occurrences in the West are concerned. But the Serbian and Russian epidemics have shown that the hero of our biography has lost none of his vigor, cruelty, and stealth, and will take prompt advantage of any relaxation of vigilance and preparedness. There is no hope that he will reform or "get religion."

Although partially and temporarily triumphant during the last war, he drew down upon himself the renewed and intensified curiosity of those who crave this kind of excitement. Not infrequently he has turned upon a pursuer and has stopped him in his tracks. But the pursuit goes on. He has been traced to all corners of the world and we know — almost, though not yet completely — where his tribe is established. His hiding places in rats, fleas, and lice have been uncovered, and if there are any further ones, still unknown, they will not remain long undetected. His methods of attack are being revealed and appropriate weapons to repulse him are being forged. In this — unlike most other matters of international interest — the whole world has coöperated against the common enemy. French, Swiss, American, British, German, Brazilian, Japanese, Chinese, Russian, and Mexican investigators have worked together, cheered each other on and helped one another in friendly rivalry. To describe their work belongs to technical literature. To attempt to do so in this book would lead us into "popular science," a form of production which we detest and have endeavored to avoid.

Typhus is not dead. It will live on for centuries, and it will continue to break into the open whenever human stupidity and brutality give it a chance, as most likely they occasionally will. But its freedom of action is being restricted, and more and more it will be confined, like other savage creatures, in the zoölogical gardens of controlled diseases.